Bauxite and
Aluminum:
An Introduction to the
Economics of Nonfuel
Minerals

Bauxite and Aluminum: An Introduction to the Economics of Nonfuel Minerals

Ferdinand E. Banks
University of New South Wales
University of Uppsala

Lexington Books
D.C. Heath and Company
Lexington, Massachusetts
Toronto

Library of Congress Cataloging in Publication Data

Banks, Ferdinand E
  Bauxite and aluminum.

  1. Aluminum industry and trade. 2. Bauxite. 3. Mineral industries—
Australia. I. Title.
HD9539.A6B35   338.2'7'492   78-24632
ISBN 0-669-02771-5

Published simultaneously in Canada

Printed in the United States of America

International Standard Book Number: 0-669-02771-5

Library of Congress Catalog Card Number: 78-24632

For Thomas 'Pocke' Banks

Contents

List of Figures

List of Tables

Preface

This is both a textbook and a reference book in the economics of nonfuel minerals. Ideally, it should be used after an introductory course in economics, but a large part of the book and all of chapter 8, which is a survey of the world mineral economy, presuppose no formal training in economics. Also included is a reasonable amount of material about the Australian mining industry, given its importance to the Australian and the world economies.

I would like to thank the Reserve Bank of Australia for granting me their professorial fellowship in economic policy for 1978, and particularly Drs. Don Stammer and Bob Hawkins of the bank for their interest in and valuable criticism of my work. I owe a debt of gratitude to the University of New South Wales for giving me a visiting professorship in the school of economics, and I would also like to extend this debt to include the students in the courses I taught in mathematical economics. They provided me with a hassle-free environment in which to think about the great world of bauxite and aluminum, and perhaps a few other things.

Next I would like to mention the enormous assistance I received from my colleagues at the university and elsewhere. In particular, I can mention Mike Folie and Tom Mozina, whose ideas I have liberally appropriated for my work, and also Don Barnett of Macquarie University, Sydney, who graciously supplied me with a stenciled copy of his forthcoming book. I was also fortunate in being able to discuss a few points concerning price systems and uncertainty with Paul Rayment of the Economic Commission for Europe and the matter of commodity stockpiles with Jay Colebrook of UNCTAD.

I must also express my thanks to Kraaft Holtz of Eurofinance and Euroeconomics, Paris, who gave me some invaluable information on certain aspects of the aluminum industry. I received valuable help and great encouragement from John Cuddy of UNCTAD who attended my recent lecture at the Graduate School of International Studies, Geneva, and who has taken a gratifying interest in my earlier books. As usual, I owe an enormous amount to the outstanding economists of the World Bank, in particular Shamsher Singh and Kenji Takeuchi. And finally, I would like to thank various colleagues in Sweden for insisting that a textbook on nonfuel minerals would be invaluable at the present time.

Bauxite and
Aluminum:
An Introduction to the
Economics of Nonfuel
Minerals

1

Background and Introductory Survey

This is a book in applied economics. Its purpose is to introduce the reader—whether he or she is an economist, engineer, worker, student, or interested layperson—to the economics of nonfuel minerals in general and the bauxite-alumina-aluminum industry in particular. Thus this book is suitable text material or collateral reading for all types of courses in the economics of natural resources or the economics of industrial raw materials. In addition, the book provides a complete and up-to-date analysis of the outlook for and problems of this industry in Australia, which presently is the largest supplier of bauxite in the world. The elementary reader, that is, the reader with no background at all in economics, will find a brief but fairly complete survey of the world mineral economy in chapter 8, which can be read independently of the rest of the book.

Note that nearly all the technical material in this book has been placed in chapter appendixes. These and other technical presentations can be bypassed with a clear conscience by all except those readers interested in quantitative economics or speculation of an abstract or quasi-abstract nature. However, the main body of the book introduces some important but simple aspects of price and production theory, which, while requiring a comprehension of secondary-school algebra, would not overstress the concentration of most people and certainly no one who has had a course or two in elementary economic theory.

To begin, we will take a brief look at the "history" and uses of bauxite and aluminum. Although aluminum was not separated out as a metal until 1825, various bauxite-type silicates were being treated as early as 5300 BC in Northern Iraq for making pottery. Exposing various clays to the hot sun or placing them next to a fire would eventually make them as hard as stone. In time some people suspected that these clays contained a metal. Early in the nineteenth century Sir Humphrey Davy named the substance "aluminum"; however, actual extraction of the metal from the clay eluded his genius.

In 1821 large quantities of an ore subsequently named "bauxite" were discovered at Les Baux, France; and in 1825 the Dane Oersted produced the metal aluminum in his laboratory. Frederich Wohler of Germany duplicated Oersted's results in 1845, obtaining in the course of his work some aluminum particles as large as pin heads. Nine years later the French scientist Sainte-Claire Deville was able to obtain aluminum lumps about the size of marbles.

As far as it can be estimated, the money price of aluminum around the year 1852 was $1200 per kilogram. With Sainte-Claire Deville's discovery, the price came down to $598 per kilogram, and production on a commercial scale, beginning in 1858, drove the price down to about $25 per kilogram. The price was $17 in 1886 when almost at the same time Charles Hall of the United States and Paul Heroult of France discovered a low-cost method of producing aluminum from aluminum oxide. Their discovery prompted a further decline in the price of aluminum. Indeed, two years later Karl Bayer of Germany developed a comparatively efficient method for extracting aluminum oxide—or "alumina"—from bauxite, and the price of the metal reached $11.55. At this point it should be noted that the Bayer process treats bauxite chemically to form alumina. On the average, the production of 2 kg of alumina requires 4 to 6 kg of bauxite. Other inputs are 0.5 kg of coal, 0.25 kg of fuel oil, 0.5 kg of soda, 0.13 kg of lime, and certain other ingredients that will be identified later. Similarly, a typical Hall-Heroult process requires 2 kg of alumina (and a number of other inputs) to produce 1 kg of aluminum.

Toward the end of the nineteenth century, aluminum entered its modern era. By 1895 when the Hall-Heroult and Bayer processes were being employed on a commercial scale, the price of aluminum was $3.2 per kilogram. In 1950 it could be purchased for $1.8 per kilogram, and at the present time its *money price* is $1.2 per kilogram. The phenomenon that the reader should be aware of here is the continuous fall in the *real* price of aluminum over the years, where the real price is usually defined as the money price divided by some consumer or industrial price index. Thus if the money price of a commodity were to double while the consumer price index increased by a factor of 4, the real price of the commodity would be halved. As a result, it would take twice as much of the commodity to purchase the same amount of consumer goods as were being purchased before the change in prices or, inversely, one half as many consumer goods could purchase the original amount of the commodity. It also appears that the real price of bauxite has fallen even more rapidly than that of aluminum.

Concerning the uses of these substances, bauxite is the basic raw material for alumina, but some bauxite is used in the manufacture of abrasives, as a catalyst in the processing of crude oil, and in the production of insulating materials, refractories, water- and sewage-treatment equipment, and so on. Analogously, alumina is the principal raw-material input of aluminum, but about 6% of the global output of alumina goes into the production of abrasives, chemical filters, and artificial sapphires.

This summary can be concluded by noting that aluminum is light, has a high strength-to-weight ratio, high thermal conductivity, and 62% of the electrical conductivity of copper. It is also nontoxic, very malleable, and nonmagnetic. Because of its strength, it can be used in many ways in the home and the industrial-construction sector where about 25% of aluminum output is directed. Worldwide, approximately 15% of the aluminum produced every year is consumed in the electrical and communications industry, 11% in consumer durables, 14% in the manufacture of industrial and agricultural machinery, and another 11% in containers and packaging. Similarly, fairly large amounts go to the manufacture of motor-vehicles, aircraft, irrigation equipment, military supplies, and metallurgical and chemical materials.

Aluminum and Its Ores

Along with iron and copper, aluminum ranks at the top of the list of *metals* insofar as its importance in industrial processes is concerned and in its value on world markets. The main reason for its importance is not just its usability but also the sheer amount of it available in *ore* form. As pointed out earlier, aluminum begins as an ore, for the most part as bauxite; but there are many other ores out of which aluminum can be manufactured. Some of the most common are alunite, kaolinite, illite, dawsonite, and anorthosite. Unlike the ores of copper, these materials are fairly dense in the crust of the earth, and thus for obvious statistical reasons they are more often found in concentrations suitable for exploitation than copper ore. Table 1-1 compares the average concentrations of some metals in the earth's crust.

The enrichment factor is the factor by which the substance must be concentrated, by nature, so that with a given state of technology, it can be considered minable. Thus if we again consider aluminum and copper, we see that there is something on the order of 25 times the probability that we will find highly minable concentrations of aluminum bearing ores than copper in a random search (assuming that the statistical distribution of ore concentration is the same

Table 1-1
Crustal Abundance, in Metric Tons, and Average Percentage Concentration of Some Metals in the Earth's Crust in Parts per Million

Metal (Ores)	Crustal Abundance (1)	Average Concentration (2)	Lowest Average in Minable Deposits (3)	Enrichment Factor (3)/(2)
Aluminum	2.0×10^{18}	83,000	185,000	2.2
Iron	1.4×10^{18}	58,000	200,000	3.4
Manganese	31.2×10^{12}	1300	250,000	190
Chromium	2.6×10^{15}	110	230,000	2100
Zinc	2.2×10^{15}	94	35,000	370
Copper	1.5×10^{15}	63	3,500	56
Lead	2.9×10^{14}	12	40,000	3300
Silver	1.8×10^{12}	0.075	100	1330
Gold	8.4×10^{10}	0.0035	3.5	1000

Source: United States Bureau of Mines.
Note: The Crustal Abundance is the *total* quantity of the resource found in the earth's crust. It might be called the ultimate amount of the resource.

for aluminum and copper). The importance of the enrichment concept can be seen from the fact that if a metal such as tin were distributed *evenly* throughout the crust of the earth, it would be necessary to remove 60 tons of rock and dirt in order to get a single kilogram of metal. With the passing of time the enrichment factor necessary for economical extraction has decreased, partially because the richest concentrations are used up, and partially because of technological advances. For instance, with the huge capacities of today's mining and processing installations, large and relatively lean deposits are more economical to exploit than small orebodies that are comparatively rich.

Goeller and Weinberg (1976) maintain that in the future it will be possible to obtain so much aluminum and iron ore from low grade ores that, together with such plentiful items as plastics and biological resources, the raw material basis for a real Eldorado might come into existence. This may indeed be so, but as will be pointed out later in this book, the energy requirements for such a paradise will be enormous. The opinion here, in fact, is that instead of beginning to contemplate a Garden of Eden in which all natural resource constraints on unlimited prosperity are removed, it would be better for decision makers to begin devoting a portion of their precious time to thinking about how civilized standards and conduct can be maintained over periods of extended shortages.

Table 1-2 partially catalogues the various aluminum bearing ores. It should

Table 1-2
The Most Important Aluminum-bearing Materials

Material	Chemical Description	% Alumina in 1 ton of Ore (Average)	Tons Ore Needed for 1 ton of Aluminum (Average)	Largest Deposits Found in:
Bauxite		50	4.75	South of France,
Boehmite	$AL_2O_3 \cdot 2H_2O$			Australia, Jamaica,
Diaspore	$AL_2O_3 \cdot H_2O$			Guyana, etc.
Gibbsite	$AL_2O_3 \cdot 3H_2O$			
Clay		30	6.0	Almost everywhere
Kaolinite	$AL_2O_3 2S_iO_2 2H_2O$			
Shale	Impure clay	30	6.0	Almost everywhere
Anorthosite		14	7	U.S. (especially
Albite	$NA_2O \cdot AL_2O_3 \cdot 6S_iO_2$			Georgia)
Anorthite	$C_aO \cdot AL_2O_3 \cdot 2S_iO_2$			
Alunite	$K_2SO_4 AL_2(SO_4)_3 \cdot \quad 2AL_2O_3 6H_2O$	17	7	U.S.S.R. U.S. (Wyoming, Utah)
Dawsonite	$NA_3AL(CO_3)_5 2AL(CH)_3$	30	6.0	U.S. (Colorado) China (Manchuria)

be noted that bauxite is not only the richest of the aluminum bearing ores, but it also contains the smallest percentage of impurities. The presence of impurities is crucial in determining the efficiency of the alumina-production process through which contaminants are drained off to get the "red mud" that is alumina. However, students of the aluminum industry have not failed to notice the decline in *ore grades* now characteristic of bauxite reserves. The tendency is to move from the best ores, which at present contain about 50% recoverable alumina, down to ores that contain 35–40%. In addition, future supplies of bauxite may be located in areas where the cost of transporting them to where they will receive their final processing is as great or greater than mining and preliminary processing expenditures. As a result, there is an ever-present incentive to advance the day when the copious supplies of clay, anorthosite, laterites, shales, and similar materials in the earth's crust can be utilized.

Production Theory: An Introduction

One of the most interesting topics in microeconomics is production theory, which for the most part deals with the relation between inputs and outputs. In broad outline, this branch of economics concerns the output of a particular commodity as it varies with the input of such things as capital and labor. At this point it is necessary to distinguish between two categories of inputs. The first are the so-called *primary inputs,* comprising capital, labor, and land. Of these, labor could be said to occupy a special place since it is not produced; but it must be remembered that there is an important relationship between the amount of capital available and the productivity of labor, and in addition, much technical progress is introduced in the form of new equipment or improvements to existing material. The odd member of this trio is land, which, while obviously having a vital role to play in the production of all kinds of minerals, is usually ignored in the textbooks.

On the other hand, the various categories of natural-resource and energy inputs, half-fabricates, and so on, are called *intermediate inputs.* Though extremely important, most students of economic theory have probably never encountered in their textbooks a production relationship that formally considers intermediate inputs. The reason for this omission is that in economics it is *net output,* or *value added,* that is usually found relevant. This concept involves only the primary inputs since in value terms net output or value added is defined as gross output minus the payment for intermediate inputs.

To get a better idea of this matter, the reader should consider the following simple example that has been condensed from an algebraic presentation in Banks (1977*b*). (It should also be emphasized that the purpose of this example is to illustrate some aspects of production theory rather than provide a realistic insight into the functioning of the bauxite-aluminum industries.) A superman, employing his bare hands, digs out 100 bauxite rocks a day on a plot of land that he inherited. He sells these rocks to a processor on an adjacent plot of land for $1 per rock. This processor, using only machines and labor, turns these rocks into 25 units of a valuable material called "aluminum." Let us now assume that these 25 units of aluminum sell for $1100.

On the basis of these data, we can examine the exact relationship between outputs and inputs. For the superman, the input is muscular exertion and the physical output is 100 bauxite rocks. The output in value terms is $100, and since there were no intermediate inputs, this is also the value added. In this case the $100 can be regarded as the return to the primary factor or, equivalently, the *wage* or *rent* of labor. (In some situations we might want to stretch this illustration a little further and call the aforementioned $100 the return to the primary factor *plus* economic profit. *Economic profit* would arise here if the superman is gaining a larger monetary reward digging bauxite than he would realize in an alternative occupation.)

On the other hand, bauxite is an intermediate good in the aluminum industry. We have taken the output of this industry in value terms to be $1100; but since $100 was paid for the intermediate good bauxite, the net output, or value added, amounts to $1000. This sum is also equal to the payments to the primary factors used to produce aluminum, which in this case would be the total of the wages (or rents) paid to labor and the rents paid capital owners. The concept of profit will be bypassed for the time being, but it can be mentioned that as in the case cited in the previous example, economic profit would accrue to the owners of capital if the percentage return or *yield* on their investments in this industry exceeded the yield on a perfectly safe financial asset such as a government bond.

Leaving simple examples in favor of simple diagrams, figure 1-1 presents a sketch of inputs and processes in the production of 1 lb of aluminum. Note that three distinct activities are being displayed—bauxite-alumina-aluminum—and it can happen that these are found in three widely separate geographical locations. Conversely, it is conceptually possible that we might have a fully integrated system where bauxite is mined at one end, and nearly pure metal in a variety of shapes comes out the other.

The intricacies of these activities will be discussed later, but for the purpose of this introduction, a few general comments are in order. The production of bauxite is a fairly simple matter, assuming that a bauxite deposit of decent size and richness is available. It may be so, however, that the overall economic advantage for a country (as opposed to a firm or a group of private individuals) is increased through processing rather than just extracting ore. This has been the claim of a number of primary commodity-producing countries in the Third World for many years.

Alumina production yields considerably larger local outlays for supplies, labor, and various services than bauxite mining; and some estimates indicate that alumina provides as much as four times as much employment per ton of output

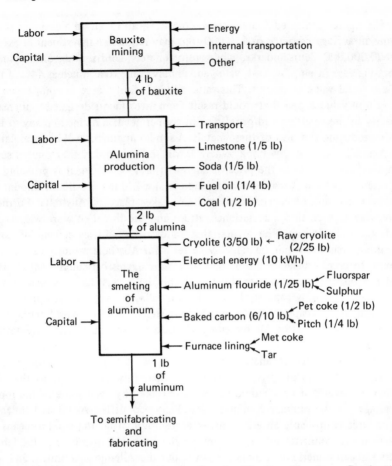

Figure 1-1. Inputs and Outputs in the Bauxite-Alumina-Aluminum Production
Cycle

Source: U.S. Geological Service.
Note: IA: Increased availability.
       IK: Increased knowledge.

as bauxite, with foreign exchange benefits from the exporting of this alumina
also reaching four times those of unprocessed ore. Similarly, countries going one
step further and producing aluminum can expect still another sizable boost in
value added in the form of an increase in total payments to domestic factors of
production. Horst Habenicht (1978) has examined this matter for Jamaica and
Surinam in an informative article and he concludes that if energy costs can be
kept down, the production of finished aluminum by LDCs may be a profitable
undertaking.

His figures indicated that in these countries, although an aluminum plant having an average capacity of 150,000 tons may require an investment in excess of 800,000,000 deutschmarks, while unfortunately contributing only a small direct increase in employment, value added may come to as much as 44% of the gross or total value of output. This value added is fairly large in comparison to the gain in value added that would result from investments designed to increase capacity in the smelting and/or refining of copper, lead, and zinc in many of the LDCs producing the ores of these metals. A similar argument will be formulated for Australia later in this book, where it will be contended that due to such things as the increase in the price of energy after 1973–1974, it is probably in the interest of both Australia and the industrial world if a maximum amount of further processing of energy-intensive items takes place in Australia. Furthermore, due to such things as nationalizations and the threat of worldwide commodity cartels, it could be argued that industrial countries consuming large amounts of bauxite would be palpably better off if Australia produced sufficient bauxite to counterbalance the designs of a more aggressive bauxite cartel. The quid pro quo for this favor might conceivably be a very large increase in the amount of processing capacity located in Australia, to include a guarantee that the output of this capacity would not be "interfered with" on world markets by the traditional processors of bauxite nor made the object of excessive tariffs and quotas.

Habenicht ends his analysis of this issue with the miners' maxim that "money is earned in the pits," by which he apparently means to imply that the further processing of ores should not be undertaken by even some of the most advantaged of the primary commodity-producing countries. As far as I am concerned, this is probably an erroneous interpretation. What the maxim means is that in many countries miners are among the best-paid members of the labor force. Chilean miners, for instance, both under the Allende government and before, were among the best paid workers in South America; and had it not been for the so-called energy crisis, the annual pay of Swedish mine workers might easily have exceeded that of Swedish professors of economics before the middle of the 1980s, which considering the utility of the work being carried out by some of these economists, might have been a good thing. However, the issue here is not the remuneration of miners and inept scholars nor, for that matter, the yield on money invested in the mining industry as compared to the "processing industry." Rather, assuming a satisfactory return on capital, it has to do with raising the level of productive employment in countries with high unemployment; increasing the payments to local as compared to foreign factors of production; and providing incentives for domestic firms to increase the quantity and quality of goods and services supplied the processing sector, thereby helping to raise the general technical and educational level.

The reader should also be aware that in 1973 the value of the world production of all metallic ores at the mining stage was $30 billion, which was less than

1% of world gross national product; while the value of the production of relatively pure metals (smelted and/or refined) came to approximately $90 billion. And these figures do not tell the entire story. The value of iron ore produced in 1973 was only a few billion dollars; but the total value of international trade in iron ore, coking coal, semifinished and finished steel products, and so on, reached almost $65 billion.

Under the circumstances, the conclusion can hardly be avoided that although mining is, for the world, an indispensable occupation, and ostensibly the "backbone of the country" for one lucky realm, there is no point in exaggerating its significance in either financial terms or as a creator of employment. At the same time it may be possible to accept the proposition that circumstances exist that militate against an expansion in further processing outside the major industrial countries. The ad valorem tariff on aluminum imports into the United States, Canada, the European Economic Community, Britain, and Japan run from an average of 8.5% on aluminum ingot and billets to more than 20% on cable, pipe, and tubes. Given these tariff levels, as well as the other discriminatory measures and restrictions placed on light manufacturers—to include smelted and refined products—destined for North America, Japan, and Western Europe, even processing facilities with an intrinsic high yield or profitability are doomed to be noncompetitive.

Reserves, Resources, and Bauxite

According to the U.S. Geological Survey and Bureau of Mines, world reserves of bauxite in 1977 came to almost 25,000 million L-T, corresponding to 5600 S-T of aluminum. With a production rate of 84 million tons of bauxite per year, which was the figure for 1977, these reserves would last 297 years; but if bauxite consumption continues to grow at the trend rate of 9% a year, these reserves will only suffice 37 years. This last figure is sometimes called the *dynamic* reserve/production ratio.

Quite apart from this, it should be appreciated that the amount of bauxite reserves is constantly being adjusted upward. World bauxite reserves were estimated at 1 billion L-T in 1945, 3 billion in 1955, 6 billion in 1965, and 25 billion in 1977. The jump between 1965 and 1977 can be accounted for by the growing attention paid Australia, Brazil, and the West Coast of Africa by various purchasers of bauxite in the major industrial countries. However, everything considered, some question must be asked as to whether there is any point in ferreting out additional reserves at the present time. Remember that a dollar invested in a bond or a bank deposit at an interest rate of 8% (which is probably typical for a long-term deposit) would yield $(1 + 0.08)^{37} = \$17.25$ after 37 years. Consequently, this amount would have to be the *profit* in money terms in 37 years in order to justify investing $1 in creating a unit of reserves that would be ex-

tracted at that time. Given the historical movement of bauxite prices, this may
be too much to ask. According to some economists, the ideal figure for the
dynamic production/reserve ratio seems to be between 12 and 15 because if it
were less, a large part of the capacity or useful life of both the extraction and
processing equipment might be wasted.

Before taking a detailed look at bauxite reserves, it should be stressed that
reserves are only a portion of the total supplies of a mineral. We must also introduce the concept of *resources,* which includes supplies that cannot be profitably
extracted given the structure of existing technology and the price of the mineral
(or, by extension, the price of the metal into which the mineral is processed). A
portion of these resources though are designated *paramarginal,* which means that
they lie on the brink of exploitability. It has been said that a 55% increase in the
price of aluminum would lead to a 40–45% increase in the stock of bauxite reserves. Equally important is the claim that a 15–25% rise in the price of aluminum would warrant the large-scale mining of aluminum-bearing materials other
than bauxite.

One of the most useful devices for discussing this subject is the classification
system developed by the U.S. Geological Survey, which is shown in figure 1–2
and which functions as follows. Along the horizontal axis are listed total supplies,
or reserves, *plus* resources. The two basic categories here are discovered and un-
discovered supplies, which are described with respect to the information that is
available about them. On the vertical axis the characteristic of interest is economic value, and here the natural partition is between profitable and unprofit-
able. Thus in the upper left-hand corner of the system, we can identify reserves
as supplies that definitely or probably exist and are profitable (to extract).

As implied earlier, resources are constantly being reclassified. On the extreme right of figure 1–2 are mineral deposits whose quantitative and qualitative

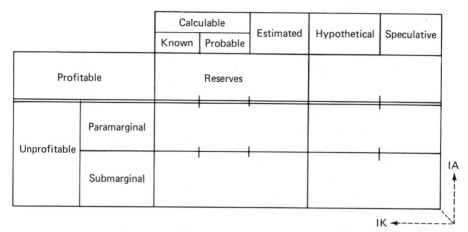

Figure 1–2. The Classification of Mineral Supplies

attributes have been assessed on the basis of a few boreholes and a study of adjacent rock formations. Still, it is a fairly widespread belief that almost any category of resource allotted a place in this system will eventually be designated a reserve. The reason for this belief is simply that given the modus operandi of the exploration services and geological surveys of the world, only prospects that are potentially profitable are examined.

It is also interesting to observe that there are concepts of mineral availability even more abstract than resources. One of these could be called "crustal abundance" or "crustal profusion" and has to do with the total resources of a mineral down to some specific depth, for instance, 1 km. Readers interested in considering this topic can refer to Banks (1976b), but in line with our present discussion, it can be pointed out that the reserves of bauxite mentioned previously (25 billion tons) are only a hundred thousandth of the known amount of bauxite in the earth's crust down to a depth of 200 m. Thus it seems likely that there are enough bauxite resources available to ensure a steady addition to bauxite reserves for a great many years, with or without substantial increases in the price of aluminum. Table 1-3 presents some information on bauxite production and reserves.

Later on this book will discuss briefly the pricing of bauxite; but as an introduction, a few remarks can be made now. Between 1968 and 1973 the price of bauxite increased by about 50%, largely because of an acceleration in the increase in demand for aluminum. In addition, during this period the price of a ton of bauxite varied between $6 and $9 per ton because of the quality differences between various bauxite deposits. In 1970 the Jamaican government insisted on doubling the bauxite price, and the foreign operating companies in Jamaica not only acceded to this demand but allowed the price increase to take place retroactively. The main reason for this unprecedented generosity was not simple altruism but rather the fact that these increased costs could be used to obtain tax deducations in their home countries.

In 1974, however, further levies by the Jamaican government on these companies were not received with such good humor. On this occasion the Jamaicans tied the price of bauxite to the price of aluminum, setting the price at which they would sell bauxite as a fraction of the U.S. price of ingot. The immediate result of this measure was a doubling of the cost of Jamaican bauxite, and moreover, the entire price rise could not be covered by tax deductions in the buyers' home countries. Just how this drama will turn out is uncertain at the present time, but there are signs that the Jamaican share of the world bauxite market has begun to weaken.

Elementary Discounting

Several topics in this chapter lend themselves to an algebraic exposition, and in keeping with the aim of this book to separate technical and descriptive material,

Table 1-3
Bauxite Capacity, Production, Reserves, and Resources (1977)

	Capacity $(10^3 S\text{-}T)$	Production $(10^3 S\text{-}T)$	Reserves $(10^6 S\text{-}T)$	Recoverable Aluminum Equivalent (million S-T)		
				Reserves	Other Resources	Total
North America						
United States	650	460	40	10	40	50
Jamaica	3800	2630	2000	450	50	500
Other	500	310	180	40	30	70
Total	4950	3400	2200	500	100	600
South America						
Brazil	300	250	2500	600	500	1100
Guyana	1100	700	1000	260	–	260
Surinam	2000	1250	490	130	110	240
Other	–	–	50	10	190	200
Total	3400	2200	4000	1000	800	1800
Europe						
France	800	460	–	–	–	–
Greece	900	670	750	170	50	220
Hungary	700	680	200	45	20	65
Italy	10	10	–	–	–	–
U.S.S.R.	1000	900	150	30	30	60
Yugoslavia	800	480	400	80	–	85
Other	200	200	90	20	50	70
Total	4410	3400	1600	350	150	500
Africa						
Cameroons	–	–	1000	200	220	420
Ghana	100	50	570	130	10	140
Guinea	3300	2000	8200	1900	–	1900
Sierra Leone	–	–	130	30	–	30
Other	200	150	100	20	300	320
Total	3600	3100	10000	2300	500	2800
Asia						
China (Mainland)	–	–	150	30	170	200
India	400	330	1400	320	80	400
Indonesia	–	–	700	150	60	210
Other	900	770	50	10	200	210
Total	1300	1100	2300	500	500	1000
Oceania						
Australia	6700	5800	4500	1000	300	1300
Other	–	–	50	10	40	50
World Total	24360	19000	25000	5600	2400	8000

Note: Bauxite capacity and production is in thousand short tons (S-T) of aluminum of aluminum equivalent. This can be converted to thousand long tons of bauxite by multiplying by 4.46. Observe also that the first column labeled "reserves" is in *million* long tons of bauxite.

these expositions can be found in the forthcoming chapter appendix. But the matter of discounting is basic to a great deal of the ensuing discussion, and in addition can be presented to readers with a minimal background in secondary school mathematics.

Our approach to discounting need involve no more than formalization of the well-known fact that a given amount of an asset is worth more today than the same amount is worth in the future, *all else remaining equal.* If we have the asset now, then we also have the option of enjoying its services if we so desire; otherwise we face uncertainties concerning our desires and appetites up to and including the date on which it is received. (Consider, for example, an automobile.) Usually if people have a choice, they demand some sort of compensation for postponing present satisfaction. Thus a sum of money (which represents generalized purchasing power) today is related to a sum in the future through a discount or interest rate that says something about the actual compensation available for someone prepared to defer a unit of consumption for a given period. This period is generally taken to be a year. For instance, if I have $100 today and the interest rate is 10% then it is customary to say that this $100 is equivalent to $100(1 + r) = 100(1 + 0.10) = \$110$ in a year's time since if I give up $100 on 15 March 1979, I can obtain a bond or bank deposit that will provide me with $110 on 15 March 1980.

It is also enlightening to turn this formulation around. In one year $110 is worth $110/(1 + r) = 110/(1 + 0.10) = \$100$ today. The general expression that we are moving toward is:

$$A_{t+1} = A_t (1 + r) \quad \text{or} \quad A_t = \frac{A_{t+1}}{(1 + r)}$$

In this expression, t signifies time period. Similarly, if we are interested in the relationship between A_t and A_{t+2}, we have:

$$A_{t+2} = A_{t+1} (1 + r) = A_t (1 + r)(1 + r) = A_t (1 + r)^2$$

If, for instance, the rate of interest is 10%, $100 today becomes $100(1 + 0.10)^2 = \$121$ in 2 years, and $21 is the premium for giving up control over $100 of present purchasing power for a period of 2 years.

We began this exposition with a brief reference to the agony of having to wait for our pleasure, and we are now talking about how money grows when used to buy a bond or is put in a savings account. The connection is roughly as follows. The interest rate is an "objective" criterion: If you visit your local bank or brokerage office, you can be quoted the interest rate on such and such a type of deposit or a security possessing a certain maturity. Let us assume that this is 10%. At the same time the typical individual will possess a subjective preference or subjective *discount rate* concerning his or her willingness to surrender money today in return for money tomorrow. Some people might regard 10% as a nice reward for waiting and thus be inclined to put a sizable portion of their monthly paycheck in a savings account offering this rate of interest; while the subjective discount rate of others might be so high that they would never consider postponing any consumption unless they were rewarded with a 100% interest rate.

Unfortunately, however, neither economists or psychologists have had much luck measuring individual discount rates and no luck at all with aggregating them. Instead, the practice has been to use interest rates as a proxy for aggregate subjective discount rates, and so these two terms are often used interchangeably, both here and throughout most of the other literature of economics. Thus a society with a low average rate of interest such as Switzerland might be regarded as a society where the average person has only a moderate preference for today's goods as compared to tomorrows.

Finally, note that if we receive various amounts of money at different times in the future, these future income flows can be discounted and summed to obtain a *present value*. As an example, we can inquire into the present value of $110 a year from now and $121 two years from now. Assuming an interest rate of 10%, we obtain:

$$PV = \frac{110}{(1+r)} + \frac{121}{(1+r)^2} = 200 \qquad r = 0.10$$

This concept can be generalized. Letting A_i represent a money flow received at i periods in the future and r the discount rate, we have as the present value of a stream of n payments:

$$PV = \frac{A_1}{(1+r)} + \frac{A_2}{(1+r)^2} + \cdots + \frac{A_j}{(1+r)^j} + \cdots + \frac{A_n}{(1+r)^n}$$

Later on we will compare payment streams, and the criterion we will adopt will reduce to ranking these streams on the basis of the size of their present value. At this point the reader should experiment with discounting and comparing various income streams with the same and different lengths, employing different rates of interest. One of the things he or she will notice is that as interest rates increase, present values decrease, assuming no change in the payment streams. This merely signifies a rise in impatience: The greater importance attached near as opposed to distant payments, and the downgrading of more remote satisfactions.

Appendix 1A
Discounting and the
Time to Exhaustion

In this and ensuing appendixes, material that I believe all readers should definitely make an effort to read is marked with an asterisk. For instance, in this appendix there are materials on discounting and present value, the effect of a positive rate of growth in the consumption of a mineral on the time to its exhaustion, and a few remarks on production theory. As far as I am concerned, only the first of these is of immediate concern to readers of this book, although the second is extremely important and definitely should be understood by the reader as a concept, even if he or she has no interest in its algebraic expression.

1A-1.* This section will extend the discussion at the end of the chapter that dealt with the present value of an income stream. Let us take a situation where we have A_0 as present income, A_1 income at the end of period 1, A_2 at the end of period 2, and so on. Assuming a constant rate of interest (or discount rate) r, we define present value as:

$$PV = A_0 + \frac{A_1}{(1+r)} + \frac{A_2}{(1+r)^2} + \cdots + \frac{A_n}{(1+r)^n} = A_0 + RA_1 + R^2 A_2 + \cdots + R^n A_n$$

where we define $R = 1/(1+r)$. If we also have $A_0 = A_1 = \cdots = A_n = A$, the preceding expression can be simplified to:

$$PV = A(1 + R + R^2 + \cdots + R^n)$$

Multiplying both sides by $(1 - R)$, we get:

$$PV(1 - R) = A(1 + R + R^2 + \cdots + R^n)(1 - R)$$
$$= A(1 + R + \cdots + R^n) - A(R + R^2 + R^3 + \cdots + R^{n+1})$$
$$= A(1 - R^{n+1})$$

Thus
$$PV = A\frac{1 - R^{n+1}}{1 - R}$$

Usually we have $0 < R \leq 1$, and thus if $n \to \infty$ we get $R^{n+1} \to 0$. In this situation the above becomes:

$$PV = \frac{A}{1 - R}$$

1A-2. The next topic concerns the effect of the growth rate of mineral use on the time to its exhaustion. If we take g as the growth rate for the mineral and X_t as the consumption of the mineral during year t we have:

$$X_t = X_0 e^{gt}$$

The term X_0 is the consumption of the resource in some arbitrary initial year. Cumulative resource use is then defined as:

$$X = \int_0^T X_t dt = \int_0^T X_0 e^{gt} dt \quad \text{or} \quad X = \frac{X_0}{g} (e^{gT} - 1)$$

With $\bar{X}$ of the resource available, we can rearrange this expression to get, as the years to exhaustion, T_e:

$$T_e = \frac{1}{g} \ln \left(\frac{g\bar{X}}{X_0} + 1 \right)$$

It is also interesting to observe the effect of changes in $\bar{X}$ on T_e. Differentiating, we obtain:

$$\frac{dT_e}{d\bar{X}} = \frac{1}{g\bar{X} + X_0}$$

What we see here is that substantial changes in $\bar{X}$ are *not* reflected in the time to exhaustion. For example, Banks (1977a) has constructed an example for the world lead supply that shows that with $g = 3\%$, an increase in the amount of reserves by a factor of 14 increases time to exhaustion by a factor of only 3.7.

1A-3. The conventional algebraic designation for the production function is $q = f(K, L)$. We define the marginal productivity of capital (MP_k) as $\partial q/\partial K$, and in a neoclassical framework, we define $MP_k > 0$. Similarly, MP_L is the marginal productivity of labor and is equal to $\partial q/\partial L$. We also take $MP_L > 0$ for all positive values of the variables. As for the wage of labor w_L—or the "rent of labor," as it is sometimes called—and the rent of capital w_K, if we take p as the price of the product, we get:

$$w_L = p\frac{\partial q}{\partial L} \quad \text{and} \quad w_K = p\frac{\partial q}{\partial K}$$

In the production function, units must be carefully specified, particularly if they are not in value terms. Thus L is in labor hours, K in machine hours, and q is in units of physical output per time period.

2 Exploration and Mining: Some Economic Aspects

This chapter will introduce two important topics: exploration and mining. In addition, building on the exposition of discounting begun in the last chapter, this chapter will introduce the theory of capital values and will offer some brief insights into the problems of mineral-exporting countries of the Third World.

As far as I am aware, no economist has yet attempted to construct a systematic theory of exploration. It cannot be denied that a few economists and geologists have attempted to model various phases of the exploration process with some simple statistical and game theoretical techniques, and it may also be the case that some day these exercises will help to increase the efficiency with which minerals are located and decisions concerning their extraction are made; but at present there is no evidence in either the textbooks or the better-known learned journals that the most skilled practitioners of analytical economics feel that there is a place for their talents in this arena.

While it is true that exploration as such is the realm of the geologist, there are a number of economic factors that intrude on every step of this work. It has become imperative for the geologist to organize his work with the intention of optimizing profit from the overall project and not just that part requiring his or her services. In other words, even preliminary investigations that indicate the presence of a rich deposit may not be followed up if they involve terrain that is not accessible by conventional means. Thus in countries like Australia and Canada, many mountainous areas are ruled out as suitable for intensive exploration just now; and in some instances the same is true for areas distant from tidewater, existing roads, or railroads.

Generally, an exploration sequence begins with a more or less superficial appraisal of an area, and with a little luck end up with the full scale development of a mineral deposit that has been found worthy of exploitation. As for the exploratory program proper, this takes the following form in the majority of cases. (1) A target area is selected on the basis of geological concepts and the information and experience at the disposal of geologists and mining engineers. (2) A geophysical survey is made which usually involves an airborne electromagnetic examination of certain areas, as well as a careful definition of anomalies. (3) Then there is the follow-up geological investigation of anomalies which may encompass electromagnetic and magnetic surveys, and geological mappings, as well as the acquisition of various properties. A positive evaluation of the evidence resulting from these surveys and mappings could result in the selection of drilling targets. (4) Finally, we have the undertaking of exploratory drilling. If this results in the actual discovery of a deposit, then it must be outlined—normally by diamond drilling—so that the size and shape of the ore body can be established, and the grade of its ores appraised. It may also happen that bulk samples are processed through a pilot plant for several months in order to complement the information obtained from drilling, and to determine the optimal metallurgical techniques to use for processing the ore.

One of the things the reader should try to appreciate is that at all stages of the above process information is being generated; and this information should be evaluated as rapidly as possible and used to develop scenarios dealing with the future development of the prospect. As soon as possible after the drilling stage has begun, for example, a preliminary design of the mine and mill should be made in order to get some idea of the profitability of exploiting the deposit; and one of the aims of outlining the deposit is to determine that portion which is to be mined in the initial 5 or 10 years of the life of the mine.

A great deal of bauxite has been found by prospecting outcrops or by tracing float or bauxite stream gravels to their sources; and deposits have also been discovered by the detection of a high alumina content in soil samples. Deposits that are very far underground can present a major problem for geologists, and usually a great deal of exploratory drilling is necessary after areas believed to contain bauxite have been selected for investigation. Once a deposit is located, it is sampled by drill holes and test pits or trenches. Initial sampling is used to determine the depth and extent of the deposit, and later on more extensive sampling is undertaken to assess ore grades. Drill holes are spaced at intervals ranging from 50 to 1000 ft., depending on the regularity of the deposit; and it should be noted that, in general, geophysical methods cannot be used to identify bauxite, although resistivity, magnetic, gravimetric, and seismic methods have been used to identify topological irregularities associated with the occurrence of bauxite.

Some Economics of Exploration

As indicated in the previous chapter, once an area has been designated a target area for geological purposes or is known to contain some category of resource or possesses some characteristic that indicates minable deposits of a mineral, then it is inevitable than it is reexamined from time to time to ascertain whether, in the light of the most up-to-date technology and evolving market conditions, its potential can be upgraded. Until recently the mineral nepheline—whose presence and location was well documented in connection with exploratory work involving other substances—was considered to be of little or no value; but now a technique has been developed for using it as a source of aluminum, and so it has been reclassified a valuable raw material. The same thing is true of taconites. Today they are an important source of iron ore, but before World War II there was no economical technique for extracting ore from these rocks. This type of phenomenon will undoubtedly continue and, as a result, counterbalance some of the more pessimistic forecasts of the depletion of mineral reserves.

With this in mind, we can turn to some data on aggregate exploratory costs. For example, Canadian expenditure on mineral exploration has increased by a factor of about 7 from 1945 to 1970. At the same time exploration costs per ton of metal produced have about doubled, while the number of ore discoveries per year appears to be on a downward trend. The latest figures reveal that the average cost of a discovery rose from about $2 million in the period 1945–1955 to about $15 million by 1976. On the other hand, the value of an average discovery went from $245 million in 1946–1954 to $711 million in 1970, thus providing some compensation for the increase in discovery costs. A worldwide survey of mining companies shows that it is not unusual for it to take up to 10 years to locate a commercially viable prospect; and in Australia a study based on 40 mines revealed that for the period 1958–1979, the average cost in 1975 dollars of making a discovery that resulted in a mine came to between $30 and $35 million. Bosson and Varon (1977) have cited the case of a major Canadian firm that spent $300 million on exploration. Over 1000 properties were examined, and of these only 7 became the sites of profitable mines.

In the case of Canada, there is no conclusive evidence that diminishing returns to exploration has set in, but as the reader can judge from the preceding exposition, there is no longer any basis for the kind of optimism that claims that the mining sector will be able to meet any and all demands placed on it over an indefinite future. In the United States, exploration has seen an increase in the number of personnel involved, as well as a very large increase in the value of the equipment put at the disposal of each of these individuals. The number of discoveries per year has not shown a long-run tendency to either increase or decrease, but the cost of finding a given amount of minerals is at least twice as much today as it was in 1955–1959, and it is increasing rapidly. The problem here, of course, is that the continental United States has been thoroughly explored, and it is hardly likely that previous efforts have missed many major deposits. This does not mean, however, that it will be impossible to raise annual production. As pointed out earlier, scientific progress should continue to function in such a manner as to increase the exploitability of many known deposits, and this includes an increasing number of those presently regarded as subeconomic.

Australia and Southern Africa (to include Namibia) may deviate somewhat from this scenario. The Australian case is particularly interesting because important discoveries are still taking place in that country from time to time, and there are many people in the Australian mining industry who seem to feel that intensifying the efforts put into exploration would result in the discovery of a great deal of mineral wealth. It seems to be true that average exploration costs in Australia are still lower than in Canada and the United States, although total exploration expenditures in Australia are accelerating both in money terms and as a percentage of the value of minerals produced. These expenditures have paralleled a rapid expansion in the supply of Australian minerals on world markets, which in turn resulted in a period of rapid economic growth and ascending levels of per-capita consumption for Australia. The extent to which this can continue, though, is uncertain since as was true with almost all the advanced market economy countries, the Australian economy began to falter at the time of the 1973–1974 energy price increases.

With aggregate world industrial growth slowing down, the demand for Australian minerals fell; and due to the ensuing fall in the income and/or profitability of some mining firms, as well as the growing reluctance of foreigners to involve themselves in Australian mining ventures during a global business-cycle downturn, there was a decline in exploratory activity. Whether all this was bad or not remains to be seen since some people have gone so far as to suggest that Australia would do well to think about ventures other than supplying the world with unprocessed raw materials. Despite a quite remarkable endowment, Australian supplies of high-grade bauxite, iron ore, and the like are not inexhaustible; and depending on the size of future generations, as well as on the intentions of the present generation in regard to the standard of living of their descendents, there may be no need for a speedier exhaustion of certain Australian wasting assets, particularly since at present it seems both politically and economically impossible to replace them with other forms of assets such as reproducible industrial capital or a more efficient educational system. Of course, it could be that given the precocious appetite of the Australian consumer for even more durables, foreign travel, and the like, there are no policy options open for any government with an oversolicitous regard for unborn as compared to existing voters.

To summarize, mining firms are usually heavily committed in terms of specialized capital and other factors of production, and thus as they begin to exhaust the core of the deposit they are exploiting, they face having to extend operations to the exploitation of less favorable ore bodies, usually at a growing cost. Since it is well known that this cost begins to accelerate as these marginal ores are depleted, the management of these firms reason that there are other reserves somewhere that can be extracted at lower costs, and efforts are initiated to locate them. Thus the issue is more than simply running out of ore; it includes ensuring a stream of revenue that is sufficient to cover the possession costs of a great deal of specialized capital.

Some question can also be raised concerning the optimal amount of exploration. It has been claimed that in the United States, an excessive, and redundant, amount of information has been generated by exploratory services working for or belonging to a large number of firms pouring over the same terrain. At the same time when and if discoveries are made, search costs, and particularly the cost of acquiring leases to drill and carry out subsequent phases of exploration and development, tend to increase rapidly. This increase occurs because reserves are usually spatially correlated, and if a minable deposit is found in a certain location, there is an appreciable chance of others being near by. Thus the seller of the leases attempts, on the basis of *incomplete information,* to appropriate a portion of the value of the deposits. Consequently, as things usually turn out, lease sellers find themselves appropriating too much or too little. (On this subject it is edifying to examine the lease-selling mechanism in the case of Alaskan oil.) Whether economic theory in its current state can solve, or even attack, problems of this type remain to be seen, but the curious reader can examine a paper by Frederick M. Peterson (1978) and some of the work produced by Brian MacKenzie for a survey of the issues involved.

Before ending this section, it should be stressed that despite the intense efforts of many mining companies to operate in an environment of certainty, or at least near certainty, they may be grappling with a problem that is, by nature, insoluble. It required 5 years and 20 million dollars of preliminary expenditures, including a host of technical surveys and reports, before a decision was made by the Bougainville Copper Co., in 1969, to invest 400 million dollars over 5 years in the mine and auxiliary facilities. (And during this period, of course, not a thimbleful of ore was shipped). Still, in retrospect, it could be argued that on the basis of actual, as opposed to expected, costs of development and operation, the mine should not have been constructed. Fortunately for Bougainville Copper, prices also developed in a manner that had not been predicted, and as a result offset these unforseen costs to the extent that, in 1973, this company earned a profit of 158 million Australian dollars. This amounted to a return of 40 percent on invested capital. But, in keeping with the thesis being developed here that, *ex ante,* one never knows how these events will deteriorate, the government of Papua New Guinea, in noting this bonanza, immediately took steps to revoke the tax holiday it had so generously granted the firm; abolished the arrangement it had entered into with the company whereby it would be allowed to write off its capital investment as soon as profits permitted (and the topic of writing off an investment is broached later in this chapter and the next); and in addition imposed an excess profits tax of 70% on all income over 87.2 million Australian dollars/year. Some people, however, have suggested that Bougainville Copper was not entirely unlucky, since had they been operating in certain other parts of the world, ownership of the firm might have changed hands almost as soon as the entries on its balance sheet were made public.

An Introduction to the Economics of Mining

Mining is an operation that features drilling, blasting, crushing, loading, hauling, and perhaps a certain amount of processing. The production factors involved are capital, labor, energy in various forms, and a number of raw materials. For classificatory purposes, we can distinguish between underground and open-pit mining. The former is an underground operation where the obtaining of ore is concerned; however, activities such as crushing and loading usually take place on the surface. Essentially, this process is expensive, and underground miners—the men who drill and blast the ore body—have become very highly paid workers in the industrial countries and also in some Third World countries.

An open-pit mine, on the other hand, is little more than a large crater scooped out of the ground by removing and carting away overburden. The exposed ore is then dug out by huge shovels and removed by trucks. The type of installation that will be employed in the case of a particular deposit depends on the depth at which ore of an economic grade is located, but on the whole open-pit mining is preferable. Due to a high degree of rationalization, open-pit methods are now allowing the exploitation of ore bodies with progressively larger ratios of overburden to mineral content, and over time a greater proportion of world mineral output is being obtained in this fashion. It has been estimated that in 1970 about 65–70% of all the ore mined in the world came from open-pit installations, while in the United States the proportion is now more than 90%. At the same time it can be noted that in the United States, the average unit value of ore mined underground was 2.4 times that taken from open-pit installations.

The stripping away of overburden usually begins immediately after the boundaries of the mine have been established and working pits designated. Given the attention now being paid to environmental considerations, overburden is often set aside and returned to the mined area to restore the land surface.

The mechanics of stripping varies from place to place. In Arkansas, where as much as 13 ft of overburden must be removed for every foot of bauxite ore exposed, draglines, scrapers, shovels, and trucks are used. In Jamaica, bauxite deposits are closer to the surface, and the overburden of vegetation and topsoil is easily removed. No blasting is necessary; shovels, draglines, and scrapers are used to load the ore; and it is transported by truck, railroad, or aerial tramline to alumina plants and ports.

At Wiepa, Australia, bulldozers and scraper loaders remove the few feet of overburden covering the bauxite deposit. Front-end loaders with a capacity of 500 tons per hour then load the ore into trucks for the short trip to a beneficiating plant where it is sized and washed. At the Del Park deposits in the Darling Range of Western Australia, the disposal of vegetation and overburden is followed by blasting the hardcap that comprises the top few feet of the bauxite deposits. Front-end loaders then load the ore into dump trucks, and it is hauled to a mobile crushing plant. Finally, bauxite is carried from the crusher to an alumina plant on a 4-mile belt system. Note that metallurgical and chemical grades of bauxite that must be transported appreciable distances are usually dried before being shipped. Since some crude bauxite may contain 10 to 30% free moisture, drying can result in a decrease in shipping costs that offsets drying costs. Surinam bauxite, for instance, is dried to 3 to 6% moisture while Jamaican bauxite is shipped with about 15% moisture.

Although costs are probably rising as fast in mining as in other industrial operations, it seems to be true that considerable economies of various types are still possible, although in the case of new deposits located in remote areas, infrastructure requirements often tend to be so great as to threaten economic viability. In examining open-pit installations across the world, we see that the output of crude ore per production worker man-hour has been on an upward trend since the early 1950s, but so for that matter have wages and capital costs. In addition, the advantages of increased mechanization are being offset to a certain extent by having to mine lower grade ores. Thus far the real costs of mining have been held down by technological progress, but whether this can continue remains to be seen. The subject of thinning ore grades, their possible effect on the cost and availability of mineral supplies, and energy problems will be taken up in chapter 7.

An important analysis of surface mining has been made by S.D. Michaelson (1974) in which he examines the productivity of some of the most important components of open-pit installations. In particular, he has looked at the drilling of blastholes, the employment of shovels and trucks, and the "stripping ratio"—which concerns the amount of material mined to send one ton of ore to the mill. Since, on the average, 45% of the cost of open-pit operations in the United States can be attributed to ore and overburden haulage, Michaelson arrives at the conclusion that ideally it is necessary to develop in-situ processing techniques that would preclude the handling of large amounts of superfluous materials; but at the same time he recognizes the near impossibility of introducing these techniques on other than a pilot scale at the present time. On the other hand, since haulage productivity is improving at less than half the rate of drilling and shovel operations, and since haulage costs may range up to 60% of operating costs in large open pit operations, he sets as a target the development of haulage systems that can operate with fewer operators and less maintenance. Some estimates indicate that if trucks could be eliminated, and shovels could be emptied directly into mobile crushers, costs would be cut by as much as one third. The problem here, though, is that hauling distances and stripping ratios seem to be increasing all the time, and thus even if larger and better haulage equipment and systems become available, it may still be impossible to press down total costs.

Although it is frequently emphasized that productivity developments in mining operations around the world are disappointing at present, particularly in view of the recent escalations in energy and labor costs, there may be some grounds for optimism. As pointed out by Michaelson, there is a 250% difference in overall productivity between mines opened in the pre-1950 period, and those opened in the early 1970s. Much of this is due to the ability of mining engineers to learn from previous mistakes, and in particular to break bottlenecks due to shortcomings in particular items of equipment. Thus, should it be possible to make some appreciable improvements in haulage systems or material, this should also reflect on the productivity of such things as shovels.

Further reflection on this topic helps explain why investment in the world mining industry has not collapsed in the face of the downturn in world economic activity following the energy price rises, and the continuing slide in the price of nonfuel minerals. The proprietors of new installations can introduce high productivity equipment *en masse,* and as a result press unit costs down under those of older installations whose rationalization and mechanization possibilities are severely constrained by their existing design. As an example of what might eventually be possible here in the long run, the *Christian Science Monitor* (April 12, 1974), has reported that Russian scientists may be active in the design of robots that can be used to extract minerals from extremely uncongenial locations. It may be that it is too early to initiate even small scale projects of this kind, but there are an increasing number of predictions that in addition to working in factories, robots could be engaged in such work as ditch digging and road laying before the year 2000; and an increasing number of Japanese scientists and engineers are engaged in the design of fully automated industrial installations. As Warren (1973) has noted, the ultimate development of underground mining would entail remote-controlled, continuous operations; and although most of us expect to see this ideal, in its fullest flowering, elude mine designers for at least another decade, the Esterhazy potash mine in Canada already incorporates a number of features that would not be out of place in the mine of the distant future.

We can now direct our attention specifically to bauxite. In Europe this mineral is usually mined in underground installations, while elsewhere open pit operations predominate. One of the most important distinctions between these methods is the ease with which the rate of production can be varied in open pit as compared to underground mines. It appears that the former display, on the average, a constant unit cost, which simply indicates that changing scale involves a proportional increase or decrease in the equipment and labor being used with existing installations. By way of contrast, anything except a small increase in production in an underground mine could require up to one year if the mine and hoisting capacity were adequate; while if an additional shaft and other expensive extensions were necessary, a construction period of at least 2 years might be required before a substantial increase in output was possible. Superficially this is analogous to increasing the amount of equipment required to expand production in open-pit facilities, except that a greater amount of time is required. But as we know from our study of elementary economics, time is money in the sense that longer construction or gestation periods translate directly into higher costs. We have seen, for instance, the rising unit costs at many coal mines in the United States as they adjusted their output in response to increases in the price of energy. These cost increases could be attributed to the undertaking of investments with long gestation periods, as well as expenses accrued in attempting to shorten gestation periods (such as outlays associated with shortening the delivery time of machinery); and also raising production without optimizing the pattern in which inputs were used. For instance, in a situation where minimum-cost production calls for a uniform percentage expansion of capital *and* labor, only labor is increased.

Where scale is concerned, the normal operating capacities of almost all of the approximately 100 known bauxite mines ranged from 80,000 to over 10,000,000 S-T per year, the exceptions being in India and Brazil where a few mines producing between 50,000 and 60,000 S-T per year provide inputs for local smelters. At present about 80% of the bauxite originating in noncentrally planned economies comes from 19 mines, where the smallest of these installations has a capacity that is slightly in excess of 500,000 S-T. In fact 15 mines had a capacity of over 1,000,000 S-T, and there are indications that this figure represents the minimum capacity required for a highly profitable mining operation.

Average production costs in the mid-1970s ranged between $5 and $18 per ton, while the cost of ocean transport was between $2 and $15 per ton, depending on the length of the haul and the state of world demand for shipping services; although the higher figure here was seldom reached. On the whole, the total delivered price of bauxite from less developed countries (LDCs) to the major importing countries (excluding certain taxes and levies) came to between $10 and $25 per ton. Assuming 5 tons of bauxite used to produce 1 ton of aluminum metal, and with aluminum prices in recent years varying from $750 to $1100 per ton, bauxite prices barely even exceeded 16% of the price of refined metal and averaged out as about 11%. Moreover, the amount of aluminum metal in a typical consumer or industrial product containing that metal is generally very small, and thus a sharp increase in the price of bauxite would have very little effect on the price of the product. Satisfactory data on bauxite mining is not easy to obtain, but the figures in table 2-1 represent the cost of bauxite to purchasers

Table 2-1
Typical Bauxite Costs (1974)[a]
(In Dollars/Tonne)

	Jamaica	Guyana	Guinea	Brazil	Australia	Average
Variable cost	3.00	3.00	3.00	3.00	4.00	3.20
Capital costs	1.07	3.20	3.20	3.20	3.20	2.77
Domestic transport	2.00	2.00	2.00	2.00	2.00	2.00
Infrastructure cost	–	2.50	2.50	2.50	2.50	2.50
Ocean transport	3.00	5.60	6.00	10.00	13.50	7.62
Subtotal	9.07	16.30	16.30	20.70	25.20	
Tonnes bauxite per tonne alumina	2.50	2.0	1.9	1.8	2.2	
Bauxite cost/tonne alumina	22.68	32.60	31.73	37.26	55.44	

Source: United Nations documents.

[a]These costs apply to just before the imposition of a production levy on bauxite by Jamaica, which increased the cost of bauxite in a tonne of alumina by $33.0. For more on this matter, see chapter 4.

in the United States in 1974, prior to the imposition of the Jamaican production levy.

In the course of this book, these costs and their theoretical background will be discussed at some length. However, it is of some interest to note that the Jamaican production levy was followed by similar production levies by most of the other producers of bauxite, with Australia being the notable exception. Thus in the world recession of 1974-1975, Jamaica's market share fell considerably (as well as its production of bauxite) while despite the worldwide decrease in the demand for aluminum, the demand for Australian bauxite rose sharply.

We close this section by making a brief examination of the topic of time scale. If we look at the Cerro Colorado copper project in Panama, where total costs in 1976 had been estimated as 800 million dollars, it was believed that the total gestation period, from technical surveys to construction of the mine and some basic processing facilities, would amount to at least 8 years. This would be broken down in the following manner.

Starting Year	Activity	Maximum Time Needed To Complete (years)
1	Technical surveys and reports	3
1	Constructing access roads	3
2	Constructing camp facilities	1-2
2	Design and construction of mine	7
3	Some infrastructure construction and development of processing plant site	2
3½	Preproduction stripping of overbody	4
"	Constructing water supply plant	4
"	Constructing tailing disposal facilities	4
"	Constructing power generation installation	4
"	Constructing auxiliary facilities	4
4	Smelter construction	3
4	Other processing facilities	3

The above would be a typical schedule encountered in "bringing in" a new copper mine; and it would be a rare project where a "greenfield" installation could be completed in less than three years, or a major expansion of an existing installation carried out in less than two.

Economic Development and Minerals Exports

To round out this portion of our discussion, we will turn now to the mining industry, mineral exports, and economic development. In the present context economic development means self-perpetuating growth in physical and educational capital and the accumulation of technical skills, particularly by LDCs. Robert E. Baldwin has written a valuable introduction to this subject.

The general theme of Baldwin's analyses is that it is natural for foreign entrepreneurs to foresake manufacturing for the exploitation of natural resources when establishing themselves in a LDC. Large-scale manufacturing operations often require skills and motivations that are in short supply in many LDCs. Then too the wage rates needed to attract labor from agriculture plus training costs are generally so high in relation to the initial efficiency of this labor as to discourage industrial operations that are not heavily subsidized by local governments. In addition, these operations tend to be risky to a quite different degree from mining and simple processing in that tastes change, and new products appear that may lead to a fatal decrease in the demand for locally produced goods, when and if they become available.

On the other hand, mining and metal-processing firms moving into a Third World country do not have to be in a hurry. Not only are they dealing with a product for which there will always be a demand but since they already carry on large operations somewhere in the world and since mining investments invariably tend to be very large, both they and their competitors have every incentive to go slow and minimize risks. In general, the practice has been to establish mining and basic processing facilities in both developed and less developed countries, which feed more sophisticated processing operations in the developed countries. This means alternative sources of feedstocks for the processing facilities, and thus makes it more difficult for the government of an LDC to interfere with the operations or policies of a mining firm. For example, in the long, drawn out aftermath of a Venezuelan government charge of misconduct against iron-ore-producing companies, new investment in that country's iron ore industry fell drastically, and Venezuela's share of world exports of iron ore dropped from 12% to less than 6% in just over 10 years. Similarly, when taxes on mining output in British Columbia and Ontario were raised, the firms involved shifted the bulk of their exploration activities outside of Canada, in particular to the United States.

But interference might be difficult for an entirely different set of conditions, because once large scale production is underway, the local government has a vested interest in maintaining the employment opportunities generated by mining and processing, and by the infrastructure investment in railroads, port facilities, and so on, which accompanies them. (There are cases, however, when governments have ignored these considerations, and without bothering to discuss the matter, opted for a bigger share of the pie—as in the previously cited case of Papua New Guinea.

As Baldwin's work has made clear, one of the principal inducements to investment in natural resource exploitation consists of the low price at which land, one of the key production inputs in any mining operation, can be purchased. (This is so because in its alternative uses, land provides a very low yield in most LDCs). Given this fact, as well as the low skill levels required for a large part of the mining labor force, particularly in the case of surface mining installations, a high yield is possible even if expensive equipment, technicians, and management must be brought in from overseas. It should also be noted that, historically, the mining industry in LDCs has yielded fairly reasonable profits to its owners—regardless of their race, creed, color, or political affiliation; and this has been shown to be the case even in high wage countries such as Australia and Canada.

The question of the transfer of technology, and the ability of mineral-producing (or "host") countries to absorb technology is also important and deserves a short comment. Evidentally, the transfer of technology has worked smoothly in almost all parts of the world, for good reasons: The mining industry is the principal source of foreign exchange for many mineral-producing countries. A great deal of their energy thus goes into thinking about this industry and its problems rather than the kind of abstract trivia that occupies so many governments in the Third World. Similarly, mining companies tend to be large, well-organized enterprises, with access to large numbers of very proficient technicians, many of whom are interested in and adept at training foreigners.

All this is very important from the point of view of providing local workers with important skills and the discipline to employ these skills. In addition, these skills and disciplines can later be taken to other industries, or even to small businesses of one type or another. Equally as important, the training of graduates is rounded out in that they can get some experience in a productive enterprise rather than the make-believe of a bureaucratic sinecure. It should also be emphasized that there are very few mines in the Third World that do not have some processing capacity attached, and the labor force and local technicians servicing these installations form the natural basis on which to progressively increase a country's engagement in such things as smelting, refining, and perhaps semifabrication. Concomitantly, given the increase in labor and energy costs in developed countries, there are many firms that are not averse to seeing a larger share of these facilities transferred to the Third World since marketing outlets remain under the control of these firms, and in many instances it increases their options when dealing with the highly paid and more aggressive employees at their installations in the developed countries.

It must be stressed though that the progression from simple mining to more complex processing and on to full-blown industrialization is not always smooth, as a brief glance at the situation in Chile over the past 30 years or so indicates. Although copper mining is a highly capital-intensive operation as compared to the local industry traditionally found in Chile, the presence of multinational companies in Chile may have hindered to some extent *the Chilean economy as a whole* from developing in a capital-intensive direction. (*Capital intensity* here refers simply to the amount of capital equipment at the disposal of each member of the working population.) The copper companies had developed certain standard techniques for doing things on the basis of problems experienced in their worldwide operations; and these standard techniques, which relied heavily on imported equipment and personnel, were applied automatically without a great deal of thought as to how they could be integrated into Chilean economic development.

At the major copper mining and processing installations, output per unit of labor and the wage rate rose steadily from the early 1920s as these installations became more capital intensive; but at the same time the share of labor payments in the value of output declined as total employment stagnated or even declined. In Britain or Germany this result would have been acceptable since it indicated the freeing of labor to move into other highly productive employments; but it was acceptable in Chile because outside the capital-rich mining sector, productivity was uniformly low.

One way out of this dilemma would have been to divert some of the capital going into the mining industry into other industries or sectors, to include agriculture. Chile, as opposed to the larger copper firms, needed less capital in mining and processing, and more in such things as the manufacture of intermediate inputs for that industry. What they got was a copper industry that became increasingly capital intensive and even less dependent on local inputs, because local manufacturers could not match the technical sophistication of imported products. It is in this sense that raising the productivity and output of a single industry can, if adequate precautions are not taken, retard instead of contribute to the development process in the long run.

Having said this, let me emphasize that I do not want to give the impression that things must work out this way, or that they will if a multinational company is somewhere in the picture. For instance, another way to get the same effect is for domestic and/or foreign capital to invest heavily in such things as breweries and distilleries, which not only are popular with consumers, but can employ fairly large numbers of individuals who might have been more useful in other occupations. As a former student of mine from the west coast of Africa once observed: "When that sort of thing takes place on a large scale, you drink your development."

Kenneth Warren (1973) claims that the initiative in the relationship between mining companies and less developed countries now belongs to the latter. Everything considered, this is probably true—although it is equally true that nobody knows for sure what the future will bring; and it also seems to be the case that developed mineral-producing countries are displaying a more pronounced aggressiveness with the mining firms operating on their soil. The government of Saskatchewan has imposed production controls on its potash producers in order to raise the price of potash; and the Australian government has, when possible, intervened in negotiations between iron ore producers and the Japanese government in order to influence the price of iron ore. Similarly, Jamaica has refused to license new bauxite-producing facilities unless alumina plants are also constructed. More striking, we see that nationalizations in the world copper industry mean that state owned copper mines now account for at least 25% of the output of copper ore in the noncommunist world, while another 12% of output comes from mines in which governments have a controlling interest.

Before concluding, it should be emphasized that many conflicts between producing countries and mining organizations can be circumvented if the ownership arrangements that characterized the world mining industry a few decades ago are modified. The Cerro Colorado copper mine is indisputably a Panamanian owned asset; but the American firm Texas-Gulf will handle the operational aspects of the project on the basis of a 15-year management contract, plus 20% of the equity; however Panama can, if it wishes, buy out this share after 20 years. Similarly, the copper mine that was being developed at Sar-Cheshmeh is entirely owned by the government of Iran; but the Anaconda Copper Company has a 14-year contract to develop and operate the mine. This commission includes ensuring that sufficient Iranians are recruited and trained to make the mine a 100% Iranian operation by the end of the contract period.

Another possibility for avoiding friction involves so called "project financing." This features mining companies committing as little capital of their own as they can to a project, and—in a sharp departure from mining company tradition—borrowing as much as possible. The logic here is that if a host country interferes with a project, it offends the maximum number of people and governments, some of whom carry more weight than even the largest mining firm. Take, for example, the Mineracoes Brasileiras Reunidas iron ore project in Brazil. Equity capital, amounting to $30 million, was supplied by a Brazilian holding company, an American mining firm, a consortium of 11 Japanese firms, a Liberian shipping firm that contracted to transport the ore, and a number of minority shareholders. The greater part of the financing, amounting to almost $160 million in debt of one kind or another, was raised from the following sources: a consortium of Japanese banks contributed 15%, the World Bank contributed 31%, the Japanese Export-Import Bank contributed 23%, and the United States Export-Import Bank contributed 17%.

Furthermore, the equity involvement in this venture can be insured, to some extent, by various government agencies like the U.S. Overseas Private Investment Corporation. Because of their official status and ability to deal with host governments diplomatically as well as commercially, they can employ a very powerful antidote against what they choose to regard as unlawful seizure: retaliation. When the Chilean government expropriated mining company assets after previously committing itself to respecting the property rights of these enterprises, a writ of attachment was granted against Chilean property in the United States, to include Chilean airliners touching down at American airfields. In addition, potential purchasers of Chilean copper were informed that they were subject to legal action if they bought the copper and paid the Chilean government rather than the copper-producing firms.

An Introduction to the Theory of Capital Values

Having looked at the elements of discounting in the previous chapter, we can begin our examination of the theory of capital values. The issue here is a fairly simple one and, at least in this chapter, makes only a few demands on the analytical powers of the reader.

We have already established that in the presence of a positive interest rate, a given number of monetary units today is worth more than the same amount at any point in the future. For instance, with a 10% interest rate, $100 today can purchase a security or savings account that will be worth $110 a year from now. Conversely, $110 a year from now has a *present value* (*PV*) of $100. Thus in a textbook world of perfect certainty, with an interest rate of 10%, you would pay $100 today for a guaranteed sum of $110 a year from now.

Still considering present values, let us consider the present value of a stream of receipts amounting to $100 a year, received at the end of the next 3 years. The $100 we get a year from now has a present value of $100/(1 + 0.1) = 90.9$. Similarly, what we get at the end of 2 years has a *PV* of $100/(1 + 0.1)^2 = 82.6$; while the *PV* of the third-year receipts is $100/(1 + 0.1)^3 = 75.1$. The total *PV* of this stream is $90.9 + 82.6 + 75.1 = 248.6$, and in a world of perfect certainty this is what a rational individual would be willing to pay for this stream. The specification of perfect certainty should be carefully noted since it ensures that anyone unsophisticated enough to pay more will be victimized by rational individuals who understand the discounting of future receipts, while these same rational individuals would be unwilling to sell this particular stream of receipts for less.

Now let us turn from financial assets such as bonds or bank accounts to physical assets such as machines. The basic problem does not change: We are still interested in the present value of a stream of receipts. But the issue is complicated somewhat by the fact that labor and other factors usually cooperate with machinery; these factors must be paid; and moreover, physical capital has a tendency to deteriorate, or depreciate, over time. In addition, taxes must usually be paid on the sales or profits resulting from the goods produced by these machines.

Before introducing depreciation and other problems, let us examine the unreal case where we have no labor or interest costs and we can buy a machine for $100 that produces goods that sell for $110 a year from now. Furthermore, assume that the machine depreciates completely after 1 year. The *cost* of the machine is $100, while *revenue* or *income* from it is $110, and so its *rate of return* in this simple case is (revenue - cost)/cost = $(110 - 100)/100 = 0.1 = 10\%$. Note that nothing has been said in this example about where the money came from to purchase this machine—that is, we did not specify whether it was borrowed or whether it "belonged" to the purchaser. Later on we will show that in a textbook market at least, it makes no difference and the origin of these funds need not be specified. However, until this simple proof has been submitted, it will generally be specified whether the money used to obtain a physical asset is borrowed or otherwise.

Next suppose that we have a rate of interest equal to 10%, and we borrow money to buy a machine that provides $120 in revenue after 1 year. Again assume that the machine depreciates completely after 1 year. In this example repaying, or amortizing, our debt requires $100. *Interest charges* are $10, which leaves $120 - $100 - $10 = $10. Discounting this $10 back to the beginning of the first period gives the amount of profit, or increase in wealth, accruing to the person or persons acquiring the machine. Arithmetically this discounted value comes to $10/(1 + 0.1) = \$9.09$. The purpose of this calculation is to measure profit at the point in time when the machine was purchased. In other words, we want to place all revenues and expenditures on a common basis, and the procedure for doing this is by considering them at the same time. Thus $120 received a year from the reference date has, under the conditions of this example, a *PV* of $109.09 on the reference date. Since we have to pay only $100 for the asset that generates this amount, our profit is $9.09, which is true regardless of whether we borrow the $100 or had it to begin with. (In the first case we end up with $9.09, and in the second $109.09.) Once again it should be emphasized that for the economist a positive profit on a physical asset means that the asset gives a larger return than would have been available on a financial asset. For example, with an interest rate of 10%, a $100 investment in a financial asset would have resulted in $110 after 1 year, instead of the $120 obtained in this example by investing in a machine.

We can now consider such things as depreciation and taxes. In the preceding discussion, we assumed that the machine simply evaporated at the end of 1 period, leaving no *scrap value.* As things stand, the tax laws in most countries specify that physical assets that have a part in creating income but that depreciate while doing so can have a certain amount of their cost of purchasing price *written off* each period as a business expense. In a world of taxes, this is an important provision. For example, consider the same 1-period machine as in the preceding example, but this time assume that operating costs for, say, wages and energy, come to $5, and assume a 25% tax rate. The net return to the purchaser of the machine would be $120 - 5 - [0.25 \cdot (120 - 5)] = 86.25$. In other words, at the start of the period, the purchaser has $100. He buys a machine and produces something from which he derives a *gross income* of $120. He pays operating costs of $5 and on a taxable income of $115, pays 25% (= 28.75) in tax. He therefore ends up with $86.5. We see immediately that he has a negative profit: Should he purchase the machine, his wealth would diminish.

But had full depreciation been allowed on the capital asset, as would normally be the case, his *taxable income* would have been $120 - $5 - $100 = $15. Applying the same tax rate of 25% means that taxes would have amounted to $3.75. The asset owners' net income at the end of the period is thus $120 - $5 - $3.75 = $111.25. Using a discount rate of 10%, the present value would equal

$111.25/(1 + 0.10) = \$101.136$. A purchase price of the asset of $100 would indicate a positive profit or an increase in the wealth of the asset holder of $101.136 - \$100 = \$1.136$. Clearly, this amount is less than the increase that would have taken place had revenue from the asset been untaxed; but as shown, it is preferable to the situation that would have existed had there not been a depreciation allowance.

We can continue this discussion with a few simple extensions. If we compare two assets, for instance, two machines, or a machine and a bond, discounting in both cases must apply to *after-tax* income. Thus with a given interest rate, a negative profit on a machine does not necessarily mean that it would be preferable to buy a bond. It may happen that there is a tax on income from bonds, and despite the negative profit, its yield is less than that of a machine. In the elementary examples presented in this section, there is a simple approach to this problem. If the income from a bond or financial instrument is taxed at the rate of 20%, then with an interest rate of 10%, a $100 bond yields $8 per year instead of $10. As a result, the effective interest rate—that is, the interest rate used in discounting—should be 8% instead of 10%. Under the circumstances, if the discounting exercise applied to the physical asset yields a negative profit, then we can categorically state that a bond, not the physical asset, should be purchased. What the reader should notice here is that apparently as the interest rate goes down, the *PV* has a tendency to go up. ("Apparently" since this effect is illusory in some cases, as will be shown in the chapter appendix.)

Some question can now be raised as to whether some market rate of interest should always be used in discounting. As noted earlier, the rate of interest differs from the *discount rate* as a concept in that the first is "objective" (due to daily quotations by banks and other financial institutions), while the discount rate is subjective in that it depends on each individuals' weighing of present against future pleasures. It may thus often happen that income streams are discounted by subjective values without the slightest relation to market interest rates. Certain hard-driving businessmen have been known to use a 20% discount rate on income streams when the rate of interest on high-class bonds or bank accounts was 7 or 8%; and when asked why, they stated that they were not interested in projects that did not guarantee a return larger than 15%. On the other hand, analysts for such projects as dams and highways and economists employed by public utilities sometimes use discount rates that are lower than the market rate of interest since they may be concerned with something other than making a profit. This important issue is examined in the following example.

Suppose we have a coal mine with two grades of coal, good and inferior. With a little care, *only* the good grade of coal need be mined, and as a result the mine can produce 1000 bundles of coal in 1 year that can be sold for a profit of $1 per bundle. This practice which in some contexts is referred to as 'high grading' shortens the life of the mine. For the purpose of this example, it will be assumed that the installation must be closed down after 1 year. But, if some of the inferior grade of coal were also extracted, the mine could last 2 years. Given the latter case, let us assume that 550 king-sized bundles per year comprising a blend of both good and inferior coal can be extracted, with each bundle again selling for a profit of $1 per year. Next let us make the assumption that employment per year is the same whether we produce 1000 normal bundles or 550 king-sized bundles. Thus if the mine functions 2 years, we have twice the employment that we would have had with a 1-year operation.

Now let us suppose that the owner of the mine is an overachiever with a deep aversion to accepting the average rate of profit. He thus decides to use a 30% discount rate when considering the two profit streams that he can obtain from his mine. The authorities, however, are dissatisfied with his decision. They say that it will shorten the life of the mine, causing employment problems and wasting the resource represented by the inferior grade of coal, which, even though inferior, is still potentially salable. They suggest he use a 10% discount rate since, according to them, a project (or *prospect*) that registers a positive profit (or an increase in wealth) when a 10% rate is used is still making a decent profit. Furthermore, to keep things simple, let us assume that the mine owner received the mine free, and thus *any* positive PV means a positive profit. Accordingly, the only problem facing this gentleman is to choose between alternative profit streams.

Figure 2-1 shows a calculation of *PV*s for the two profit streams, using both the mine owner's preferred discount rate and that of the authorities. As shown, the mine owner receives his profits at the end of the year; and discounted values—discounted, that is, to the beginning of the first year—are shown in parentheses.

As the calculation shows, a 30% discount rate rejects the 2-period scheme,

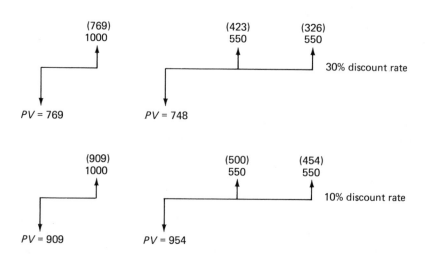

Figure 2-1. Two Profit Streams Discounted at Different Discount Rates

while a 10% discount rate accepts it in the sense that it becomes preferable to the 1-period scheme. This type of conflict has led to the use of the expressions *social* and *private* discount rates, where in terms of the preceding example, the social discount rate would be that chosen by the government of the theory that the smaller the discount rate, the more weight given the future. (On the other hand, a 30% discount rate had the effect of "cutting off" the future).

One last comment: Coal was used in the preceding example because there are a great many people who believe that where deposits of coal are concerned, the best grade of coal should be extracted first, regardless of the discount rate. In addition, the shortening of the life of the deposit through extracting the best grade first is a well-known occurrence to many coal experts. But it can also happen with other minerals that certain rates and modes of extraction cause resources to be left behind permanently. What happens is simply that the inferior grades are mixed with "waste" and become so diluted that given the present state of the mining arts, it is not worth while to attempt to recover them.

Appendix 2A
Interest and Discount Rates, and a Derivation of the Present Value Rule

2A-1. Chapter 1 suggested that there was a strong affinity between the "discount rate" and the "interest rate," and to a considerable extent, these terms would be used interchangeably in this book, as they usually are in the literature of economics.

But as mentioned, this usage is not entirely correct. The discount rate is a subjective parameter and, if applied to an individual, would say something about a person's patience or impatience where present sacrifices and future rewards are concerned. On the other hand, the *time path* of the interest rate is formed by the interaction of two factors: the same subjective discount rate referred to previously and the productivity of capital. The interest rate itself, which is usually thought of in terms of money, is ineluctably connected to the physical productivity of capital, and in a perfect market the "productivity of money" and the productivity of capital are the same.

One way to examine this topic is to look at a transformation- or production-possibility curve involving consumption in period 1 as compared to period 2. The basis of this curve is the productivity of capital: If capital is highly productive, one unit of consumption sacrificed today and the resources that would have been used to produce it used instead to produce machines for producing consumption goods would mean a stream of very many units of consumption goods, beginning at some point in the future. The transformation curve is shown in figure 2A-1 and features diminishing returns to the present sacrifice of consumption goods.

"Diminishing returns" in this context means that it becomes progressively more difficult to increase future consumption by giving up further units of present consumption. Now, the *absolute value* of the slope of the transformation curve $Z'Z''$ can be shown to be equal to $(1 + \bar{r})$, where $\bar{r}$ is the market rate of return, or the real rate of interst, and $(1 + \bar{r})$ indicates that 1 unit of consumption given up today can be transformed into $(1 + \bar{r})$ units in the following period. Note also that this slope varies as we move along the transformation curve, having its highest absolute value at Z' (where a maximum amount of consumption in the present period is available due to making no provision for the future. Thus if one unit of present consumption is sacrified, the diagram indicates that the resources used to produce that unit would find an extremely productive outlet, which may not be true for the 100th unit).

Similarly, the absolute value of the slope of either of the two sets of intertemporal utility curves in figure 2A-1(b) is equal to $(1 + n)$, where n is the

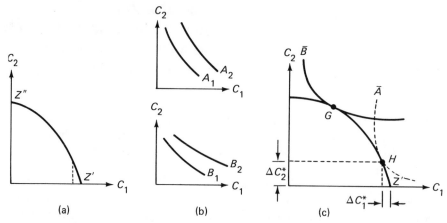

Figure 2A-1. Intertemporal Consumption Preferences and Investment Possibilities

subjective discount rate or, more correctly, the consumer(s) rate of time preference for consumption in period 1 instead of period 2. Note that the A curves are shaped in such a way that a great deal of consumption is necessary in period 2 to induce the sacrifice of a unit of consumption in period 1, while the opposite is true with the B curves. Finally, figure 2A-1(c) brings together these components of the consumption-investment situation, with specimen "equilibria" formed at G and H. What we have here is the adjusting of subjective preferences to market opportunities via the equating of time preferences between pairs of periods to the market rate of return. Note also that in line with our previous discussion, A-type curves result in an equilibrium where of the total amount of C_1 available, Z', only a small amount, ΔC_1^*, is sacrificed in favor of future consumption, ΔC_2^*. (This topic is continued in 2A-3).

2A-2.* An example presented in the last part of this chapter suggested that increasing the interest rate tends to reduce the desirability of income streams received in the distant future. This type of wording implied in the example that in order to bring into favor longer-lived projects, it was necessary to get the interest rate down. But this is not always so. For example, comparing the PV of an asset that yields 100 one period from now and 140 three periods from now; to an asset yielding 238 two periods from now, assuming that these two assets cost the same, we would get

Interest Rate	PV_1	PV_2
0%	$100 + 140 = 240$	238
100%	$\dfrac{100}{(1+1)} + \dfrac{140}{(1+1)^3} = 67.5$	$\dfrac{238}{(1+1)^2} = 59.5$

Thus we see that at both low and high interest rates we prefer, on the basis of present value, the first arrangement. What has happened is that at a low interest rate there is no discounting of the more "distant" returns, while at very high interest rates the distant returns are discounted to such an extent that they become quite small. But it happens that there is a "band" of intermediate interest rates in the vicinity of 13% where the second scheme is preferable. For example, with $r = 0.13$, we have:

$$PV_1 = \frac{100}{(1.13)} + \frac{140}{(1.13)^3} = 185.5$$

$$PV_2 = \frac{238}{(1.13)^2} = 186.3$$

Notice what has happened. With an interest rate of 100% we prefer scheme 1. Lowering the interest rate causes us to prefer (on the basis of present value) scheme 2. But further lowering the interest rate to 0% makes scheme 1 preferable again. This effect is called *reswitching*.

2A-3. On the basis of the discussion in 2A-1, and the situation at point Z' in figure 2A-1(c), it should be obvious that the rate of return, or real interest rate, can be obtained from the following expression:

$$\frac{\Delta C_2}{\Delta C_1} = \frac{\text{Return}}{\text{Cost}} = 1 + \frac{\text{Return} - \text{Cost}}{\text{Cost}} = 1 + \bar{r}$$

or

$$\bar{r} = -\left[\frac{\Delta C_2}{\Delta C_1} + 1\right]$$

But it should also be clear that if the market rate of return (or *real* interest rate) is greater than the *subjective* discount rate ($\bar{r} > n$), then the investor should continue giving up units of *present* consumption, because the *future* consumption that will result from this sacrifice will have a greater value for him. Accordingly, at point H in figure 2A-1(c), we have an equilibrium in that the real interest rate equals the subjective discount rate. In symbols, this means that $\bar{r} = n$. Thus, the criterion for investing, taking into consideration the previous manipulation, is:

$$-\left[\frac{\Delta C_2}{\Delta C_1} + 1\right] \geq n$$

Rearranging this expression we get:

$$\frac{\Delta C_2}{1 + n} \geqslant \Delta C_1$$

or

$$-\Delta C_1 + \frac{\Delta C_2}{1 + n} \geqslant 0$$

This is the present value criterion that we introduced earlier. Here we have merely derived it from elementary theoretical considerations. Once again observe that in this present value calculation, the discount rate is n (which, as pointed out earlier, is a subjective value); but for the most part in this book we will employ r, a market (and money) rate of interest. The principal reason for this is that r is well defined in that it can be obtained from a number of publications at any given time, or even a call to your local bank.

3

Processing, Secondary Metal, Market Forms, and Capital Costs

Our next topic deals with the processing of ore. As explained earlier, the processing cycle, which can extend as far as the production of semifabricated items, is preceded by mining and often a form of preliminary processing at the pit or mine for the purpose of avoiding the transport of large amounts of rock or earth containing only small amounts of the mineral. Copper typifies this arrangement, in that mining is followed by milling and concentrating, with each step of these operations raising the "purity" of the commodity and making it more economical to move.

For aluminum, processing formally begins with the production of alumina and continues with the production of aluminum metal in a smelter. Smelted aluminum is then turned into sheets, rods, and other highly usable shapes. The final step is the fabrication of a particular finished item. We often find a geographical diversity of the various sites in the mining-fabrication cycle. At present about half of all bauxite is extracted in LDCs, who also produce considerable alumina. On the other hand, the smelting of alumina into primary aluminum is an activity that takes place mostly in the industrial countries, and the same is true for the production of semifabricates. However, the most important producer of bauxite, Australia, is also responsible for a large amount of processing.

Another interesting source of aluminum is scrap aluminum, which is obtained either from the production process in such forms as shavings or in the salvage and breaking up of finished items such as cables and transport equipment. It is essential to be aware of the potential importance of scrap to the main industrial countries since scrap is a domestic source of mineral supplies that is, at least up to now, independent of international politics and the whims of this or that cartel. In almost every industrial country there is a huge *stock* of finished products that could, in an emergency, be turned into a source of aluminum, copper, zinc, and so on, provided that the secondary processing capacity existed. Metals won from these *recycled* finished products are called *secondary metals* (as opposed to the *primary* metals that have their origin in virgin ore).

Occasionally it is claimed that with primary producers becoming more involved with scrap, though not the other way around, the distinction between the two industries is becoming less clear. As the reader will soon note, technical considerations make it quite unlikely that anyone seriously intent on understanding the world aluminum market will be able to dispense with the terms "primary" and "secondary." Similarly, these industries are still keen competitors, and more important, they display sharp differences in both market and ownership structures.

The markets for primary metals tend to be highly oligopolistic. For example, six multinational giants control 60% of the world capacity of aluminum. There are a number of minor players in this marketplace, some of whom can be considered highly efficient producers and aggressive sellers; but not only do they lack the broad range of products of industry leaders like Alcoa and Reynolds but they also suffer important disadvantages where, for example, financial resources, connections (in both high and low places), and depth of experience are concerned. They sometimes find it difficult to hire the most high-powered (and expensive) technical expertise and lag even further behind in recruiting young people graduating from technical schools and universities. Many of these smaller firms are often unsure if they will have sufficient access to raw materials and intermediate products to ensure full production runs while an organization like Alcoa—the most important producer—has its own bauxite mines, produces a part of its electricity requirements, and even goes so far as to manufacture its own pots and pans. But, as pointed out later, the entrance barriers to this industry can be hurdled by certain types of intruders.

Secondary producers are usually small- to medium-sized firms that function as price takers in a highly competitive market. The secondary-processing industry has also tended to be labor intensive in comparison to the capital-intensive operations that typify the primary-processing industry, and thus it has been caused some problems by the rapid increase in the price of labor relative to the price of capital that has characterized the last five years or so. However, some compensation has been realized of late due to the effect of the escalation of energy prices on the energy-intensive primary aluminum industry. Before continuing, the reader should reexamine figure 1-1, which shows the throughput of metal from mining to semifabrication for the primary aluminum industry.

Of particular importance is the fact that the primary industry can process bauxite all the way to finished products or, with a smaller amount of processing, sell such products as sheets and ingots to independent fabricators and foundaries. Furthermore, since such processes as casting traditionally belong in the secondary cycle, there is often a movement of aluminum from the primary industry to the secondary. A similar observation is true for scrap since most of the feedstock for the secondary industry comes from the milling and fabricating operations of primary industry where a large amount of new, or *prompt,* scrap is generated.

Alumina and Aluminum Metal

The background of the alumina-production process was sketched earlier, and the reader interested in its details is advised to consult a technical reference; but some economic questions associated with the production of alumina are of interest to us at the present time.

Installations using the Bayer process for producing alumina tend to be extremely large, reflecting the immense capital requirements of the process. The smallest profitable plant for producing alumina should have a throughput of 300,000 tons per years; and installations are now being designed with a much larger capacity. On the average, about 40% of the production costs of alumina are capital costs; and in examining these costs on a global basis, it appears that Australia is the lowest cost producer. The cost of alumina at the Australian Gladstone plant averaged about $66/ton in 1974, while in the same year alumina produced in a U.S. gulf coast plant from Caribbean bauxite ranged up to $110. According to D.W. Barnett (1977), however, Australia has reaped little of the advantage of these low costs since the plants producing this alumina are owned largely by foreign interests who ship much of the alumina abroad for conversion to aluminum in their own smelters.

Whether this situation will exist in the future remains to be seen. With the international scene characterized by rising production costs and uncertainties concerning the future availability of energy and in light of the growing emphasis on environmental considerations, there are increasing chances that a great deal of processing capacity will soon be shifted to energy-rich, sparsely populated natural-resource producers like Australia, or LDCs with low wage costs that are close to major markets. Since 1965 the average cost of electrical power has increased rapidly in the United States and should continue to do so; and although existing aluminum smelters in that country cover about one third of their present power demand under contracts that are not due for renewal until 1983, the assumption is that many smelters cannot avoid a situation in which they will have to compete for a limited supply of energy. By the same token, the Japanese are facing an even more bleak predicament; in the last four or five years their aluminum industry has sustained losses that amount to at least three quarters of a billion dollars. At present, many of the major aluminum firms with financial and managerial interests in the Australian aluminum industry are actively promoting the expansion of smelter facilities. In fact, if all mooted projects are realized, present Australian smelter capacity of about 270,000 tons of aluminum would more than double by 1985. Table 3-1 provides some information on world alumina and aluminum capacity and production in 1977.

Primary Aluminum and Semifabricates

Aluminum is separated from its oxide by electrolysis, with the basic process amounting to pure alumina being reduced in a bath of fused cryolite at a temperature of approximately 1000°C. Eventually aluminum settles on the bottom of the bath (which is constrained by huge bathtub-like containers called *electrolytic cells*), where it is tapped, alloyed (if desired), cast into ingots, and cooled.

The preceding process takes about four hours, and during this period about

Table 3-1

World Alumina and Aluminum Capacity and Estimated Production (1977)

(*Thousands of Short Tons of Alumina or Aluminum Equivalent[a]*)

	Alumina		Aluminum	
	Capacity	*Production*	*Capacity*	*Production*
North America				
United States	4066	3450	5193	4539
Canada	723	650	1175	985
Jamaica	1750	1170	–	–
Other	–	–	50	46
Total	6539	5270	6418	5570
South America				
Brazil	225	150	198	185
Guyana	203	150	–	–
Surinam	776	700	73	64
Other	–	–	254	96
Total	1204	1000	525	345
Europe				
France	750	620	452	439
West Germany	993	800	841	818
Greece	287	260	160	138
Hungary	443	440	83	78
Italy	528	480	321	283
Norway	–	–	765	698
U.S.S.R.	1820	1500	2695	2425
Yugoslavia	520	310	198	218
Other	456	400	1826	1691
Total	5797	4810	7341	6788
Africa				
Ghana	–	–	220	160
Guinea	402	320	–	–
Other	–	–	267	233
Total	402	320	487	393
Asia				
India	390	250	385	197
Japan	1513	1000	1755	1325
Other	455	420	635	470
Australasia				
Australia	3871	3830	275	267
New Zealand	–	–	165	160
World total	20171	16900	17986	15515

Source: U.S. Bureau of Mines and Department of the Interior.

[a] "Aluminum equivalent" is the amount of aluminum that could be produced from the actual alumina tonnage.

16,000 kWh of electricity per ton of aluminum is required, although several thousand kilowatthours are shaved off this figure in the most modern plants. Energy costs are therefore considerable and may in fact be rising, but the physical input of energy has shown a marked tendency to decrease over time. Forty years ago it required 52 hours and more than 30,000 kWh to obtain a ton of aluminum. Until recently the target rate for energy usage appeared to be about 9000 kWh, but as will be pointed out in chapter 7, pilot processes may already be available that require much less electricity.

As shown in table 3-1, the major western producers of aluminum are the United States, Japan, Canada, Norway, and Germany. These countries account for 55% of present-day supplies and together with the U.S.S.R., produce 70% of the total world output of this metal. Australia should be judged a major potential producer; and because of its large supply of high-grade coal and other energy resources, it is on the way to reversing the situation existing prior to 1973–1974 when it was less expensive to export bauxite and alumina to Japan than to process it into metal in Australia. Canada and Norway have become important producers of aluminum ingots because of the availability of electricity in these countries; and due to the oil and gas supplies of the North Sea, both the United Kingdom and the Netherlands are making plans to expand primary production. Other countries with an important energy potential that are in the process of setting up aluminum-production facilities include Iceland, Bahrain, Mexico, Surinam, Venezuela, and Ghana.

Going one step further in the processing cycle, we see that seven major western industrial countries account for 90% of noncentrally planned semifabricate production, while the leading minor producers are Belgium, the Netherlands, Norway, Greece, and perhaps India. The story here is that semifabrication has a tendency to take place close to the final consumer since transporting the awkward shapes that characterized some semifabricated products can be expensive. Perhaps more important, import tariffs can be quite high on fabricated or semifabricated products as compared with raw materials. Among the reasons for these high tariffs is the desire by producers of these products to protect what have traditionally been very profitable activities. In fact, in the 1950s at least one major producer, Reynolds Aluminum, regarded fabrication as the profit center of its operations. Later one, however, the extraction of bauxite regained this position.

Secondary Materials

Secondary materials, or scrap, have already been introduced in this book but there are still a few aspects of this important topic that should be surveyed. One of the main attractions of the secondary aluminum industry is that, on the average, the energy used to produce a unit of aluminum from secondary materials is between 5% and 20% of that needed to produce an equivalent amount of primary aluminum; and in some uses secondary aluminum is for all practical purposes interchangeable with primary aluminum. Almost all the 3 million automotive pistons produced annually in Australia are made from recycled aluminum; and secondary materials play a major role in the production of, for example, automotive cylinder heads and garden furniture.

In addition, investment costs are lower for secondary processing facilities than for primary. In the United States, for example, the cost of the former are about 15% under those for a standard Hall-Herault installation handling the same capacity. Where the price of secondary aluminum is concerned, it tends to be somewhat lower than the price of primary metal, which simply reflects the absence of complete interchangeability between the two materials. Price *movements,* on the other hand, are similar in so far as directions are concerned, although secondary prices (and production) are much more volatile over the business cycle. For instance, total United States scrap production increased from 4.9 to 6.0 million tons between 1972 and 1974, and then fell to 4.5 million tons in 1975. Obviously, no component of the primary market has experienced such fluctuations in output. At the same time the price per ton of scrap (in dollars) was $37 in 1972, $58 in 1973, $108.52 in 1974, and $72.75 in 1975. These figures undoubtedly say something about the degree of competition in the secondary processing industry. Table 3-2 shows the production of secondary materials in the principal industrial countries.

Table 3-2
World Production of Secondary Aluminum (1971)
(*Thousands of Tons*)

Country	Production	Percent of World Production	Percent of Country's Total Production of Aluminum
United States	739	30	17
Japan	349	14	28
West Germany	279	11	–
United Kingdom	182	7	–
Italy	150	6	–
France	97	4	–
Canada	34	1	–
Spain	29	1	–
Sweden	20	0.8	21
Australia	18	0.7	8
Switzerland	14	0.6	13
Brazil	12	0.5	–
Others	38	0.15	–
World total	2467		

Source: F.E. Banks, *The Economics of Natural Resources* (New York: Plenum, 1976, and *Scarcity, Energy, and Economic Progress* (Lexington, Mass.: Lexington Books, D.C. Heath, 1977). Also World Banks and OECD documents.

The Simple Analytics of Recycling

Next we can turn to a simple algebraic delineation of this topic. It could be argued that essentially there is a direct proportional relationship between the output of a product and the demand for the inputs to be used in that product. Mathematically this reduces to an elementary expression such as $X_i = aQ_i$, where X and Q are input and output, respectively, and a is a constant of proportionality known as an "input-output" coefficient. As a result, if as is historically the situation, output increases yearly, so does the demand for inputs; but on the other hand, the amount of product that is the object of the scrapping and recycling is related to the lower aggregate production levels of previous years. For example, let us assume that only automobiles are suitable for scrapping, that the "life" of an automobile is 10 years, and that planned automobile sales for the present year are 5 million vehicles, as compared to only 1 million 10 years ago. Also assume no design changes for these cars over the 10 years. We then see that even if we could scrap and recover every nut and bolt in those 10-year-old vehicles and recycle them as inputs into current production, scrap would still account for only a fifth of the inputs needed for this year's output.

Moreover, if as is usually the case, only a fraction of the potentially recyclable materials can be recovered, it would not be possible to construct even a fifth of the present year's output (or 1 million vehicles) from recycled materials. This leads to an algebraic exercise, which we begin with some notation. First we represent industrial activity in the year t by the variable Y_t, where by industrial activity we could mean industrial procuction or gross national product. From our previous discussion we then have $C_t = \theta Y_t$, where C_t is the consumption of the input, and θ the factor of proportionality we called an "input-output" coefficient." Taking g as the growth rate of Y, L the average durability of goods containing the particular input, and λ the fraction of the input that can be recovered from the products in which it is used, we have:

$$C_t = \theta Y_t = \theta Y_0 e^{gt} \qquad (3.1.)$$

This is the amount of the input used in year t. With durability L, the amount available for recycling is that used L years earlier, or:

$$C_{t-L} = \theta Y_t e^{g(t-L)} \qquad (3.2)$$

With a fraction λ of this recoverable, we have as the amount recovered:

$$R_t = \lambda \theta Y_0 e^{g(t-L)} \qquad (3.3)$$

Thus the ratio of the amount recovered to the total consumption of the input is:

$$F = \frac{R_t}{C_t} = \frac{\lambda \theta Y_0 e^{g(t-L)}}{\theta Y_0 e^{gt}} = \frac{\lambda}{e^{gL}} \qquad (3.4)$$

It should be noted that this is an *equilibrium* value associated with the long-run constant growth of an economy. We can now examine this equation in light of some information we have on secondary materials, although we shall be concerned principally with old scrap. A large part of the data in table 3–3 is derived from my earlier work, in particular (1976b) and (1977b), but the reader should also refer to the work of Professor David Pearce (1976).

Estimating the total amount of potentially recoverable scrap is not an easy job, but the Commodity Research Unit, an especially knowledgeable London-based consulting organization, has reckoned potentially available scrap from copper in use in the United States, United Kingdom, France, and West Germany as 33 times present consumption levels. A figure of this magnitude indicates that over a limited period, it is theoretically possible for some countries to operate their economies at normal intensity employing only secondary materials as a source of metal inputs. It is doubtful, however, if the situation envisaged by Professor Glenn Seaborg (1974) and others could be brought about in which secondary materials become the chief source of industrial feedstocks, unless a no-growth or negative-growth situation were acceptable. Still, impressive economic and social gains might be possible if the scale of recycling could be stepped up in the industrial countries. In 1975 in the German Democratic Republic (GDR), almost 517,000 tons of used paper was collected, which is equiva-

Table 3–3
Some Recycling Data for Five Metals

	Old and New Scrap % (Share In)		Old Scrap (Share In)				
	Market Economies	U.S.	U.S.	g(%)	λ	L	F^b
Copper	40	46	20	4	0.75	30	0.23
Aluminum	20	18	5	9	0.75	30	0.05
Lead[a]	46	49	39	3	0.55	8	0.43
Steel-Iron	32	41	37	–	–	–	–
Zinc	21	26	5	–	–	–	–

[a]The short lifetime of lead products is principally due to the use of lead in batteries having a short "life." Similarly, its low recoverability can be traced to the widespread use of this metal as an additive in gasoline and the manufacture of paints, which prohibits recovery.
[b]Calculated from equation 3.4.

lent to paper from 4 million trees. A target of 345 million bottles and 190
million tin cans was also announced, which, among other things, would represent
the saving of 90 million kWh of energy. It seems likely that when savings of this
magnitude are possible, they justify the establishment of a comprehensive recycling industry with perhaps thousands of employees, many of whom might otherwise be unemployed. (See chapter 8 for a description of a U.S. installation
designed for "total recycling."

Aluminum and Market Forms: An Introduction

The primary aluminum industry is an oligopoly in the sense that an oligopoly is
usually defined as an industry in which a small number of firms produce the
bulk of industry output. (By way of contrast, the *secondary* aluminum market
exhibits many of the characteristics of pure competition since there are a large
number of scrap collecting and processing firms, each of which produces so small
a proportion of total output as to have little or no effect on the market price.)
   A problem arises, however, when we attempt to associate this particular
industry with the conventional textbook models of oligopoly. The basis of market power in the great world of aluminum has traditionally been the ownership
or control of bauxite mines, and until recently four firms—Alcoa, Alcan, Kaiser,
and Reynolds—controlled 65% of bauxite-mining capacity, and they used this
capacity to enjoy the prerogatives associated with supplying world markets with
a large portion of one of the most important industrial inputs. But even before
nationalizations and nationalism weakened the access of these major producers
to their supplies of bauxite, other firms with only a limited access to ore were
able to establish themselves in this industry and maintain their tenure during a
period when, because of general overcapacity, the management of the leading
firms would have been quite happy to see them disappear.
   If a long-run trend were visible at the present time, it would point toward a
weakening of the market power of the major producers. There are a number of
reasons for this trend, beginning with the postwar redistributions in world political power and continuing to the present redistributions of both political and
economic power—mostly to the disadvantage of countries like the United States
and Britain. But there have also been important judicial pressures on the market
leaders. In a famous decision Judge Learned Hand concluded that Alcoa was
socially dangerous because of its size and as a result placed limitations on both
its size and market share that resulted in the formation of three new companies:
Kaiser, Reynolds, and Alcan. This decision attracted a great deal of attention in
academic economic circles, and almost all observers are in agreement that from
the point of view of resource allocation, there was little to gain by permitting
Alcoa to literally control the supply side of the aluminum market. But by the
same token, many felt that some of the arguments advanced by Judge Hand

hardly deserve to be labeled nonsense. For example, he was able to convince
both himself and a number of infuential people that Alcoa, via its corporate
planning, was able not only to control the existing market price of primary
aluminum but also the price and supply of secondary aluminum over a fairly
long time horizon.

At the present time many economists are watching with increasing interest
the progress of a firm called AMAX (American Metal Climax) in the aluminum
market. This organization has gained access to important bauxite supplies in
Australia, particularly in the Mitchell Plateau and the Kimberly ranges of Western Australia, and is in the process of forming consortia with Japanese and
European firms under the designation ALUMAX. Although on the surface it
appears that AMAX's bauxite connection in a politically dependable country is
the key to the good fortune it has enjoyed so far, an equally important factor is
its willingness to work in close cooperation with Japanese capital, connections,
and know-how; and to serve as a link between Japanese geoeconomic ambitions
and the American market. Then too with the world aluminum market in a transi-
tion state because of the energy situation, and a likely fall in the trend growth
rate of consumption in North America and Western Europe and potential rises
elsewhere, it seems very probable that global production capacity will have to be
rearranged to a certain extent.

The upshot of the previous remarks is that despite a real short-term influence
over aluminum prices, existing aluminum producers cannot effectively exclude
everyone desiring to join their club. Aggressive firms with large financial and
technical resources who are ready to resort to international alliances can establish themselves in this industry and, depending on market conditions, prosper.
Of course, some question would have to be raised about the chances of their
success if they aimed at taking a really large share of the market. After all, aluminum producers cannot legislate demand, and this market is not expanding so fast
that it could absorb a substantial new increment of production without placing
a sharp downward pressure on the price of aluminum metal.

Where price policy is concerned, the aluminum industry has been studied in
great detail, and it appears that the normal mark-up over cost has ranged between 10 and 20%. This mark-up has generally amounted to between 2 and 4¢
per pound, with the major exception being the period right after the oil price
rises of 1973–1974 when the frenetic buying of misguided speculators drove the
price of primary aluminum sky high. As for the profit rate, which for our purposes can be defined as net after tax income divided by the value of stockholders'
equity, very little distinguishes the profit rates of the oligopolistic firms in the
aluminum industry from those of firms participating in so-called competitive
markets. The average profit rate over the postwar period for the "big six"
came to 11–12%, and for the most part this profit rate was marginally worse
than the return to durable-goods manufacturers whose oligopolistic traits are

much less distinct than those of Alcoa or Kaiser. The opinion here, in fact, is that there is precious little evidence that the major aluminum firms are still in position to victimize the public through their size or pricing policies. Moreover, given that capital requirements in this industry are so large and scale economies so important, it might be best for all concerned if the economists and lawyers intent on breaking up the large aluminum firms into smaller units would devote their valuable energies to rectifying some of the more flagrant social deficiencies of the societies in which they live.

The Price of Capital Services

Our next job is to examine one of the costs of producing an important commodity such as aluminum. Since labor costs and energy costs have a straightforward interpretation, the one we choose here is "the price of capital services" or, as it is sometimes called, the "capital cost."

My intention is to approach this matter with a simple numerical example. I would strongly advise, however, all readers to work through this example until they understand it perfectly since it contains some extremely important economic concepts. To begin, let us assume that you borrow $1000 to buy a physical asset such as a machine or house. For simplicity, we will assume that the asset has a "life" of 2 years, after which it falls apart. We shall also assume that you contract to repay the loan in 2 years, and the rate of interest is 10%.

Next let us examine this situation from the point of view of the lender. He or she has $1000, which if lent out at a 10% interest rate, will bring $1000 $(1 + 0.10)^2 = \$1210$ at the end of 2 years. Now in a perfect (that is, textbook) market, the lender is interested only in getting $1210 in return for giving up $1000 two years earlier. But some skeptics claim that the real world is not perfect; and these people, among others, have devised schemes whereby people who borrow money are compelled to pay it back periodically rather than in one sum at the time the loan becomes due. Later on we shall break down these periodic payments into *amortization,* or *depreciation,* payments, and *interest* payments; but to begin, we will simply lump them together. Our period will be 1 year.

At the end of the first and second year, the borrower makes a payment to the lender. These payments are of such a magnitude that the lender is indifferent between these payments and the certainty of the sum of $1210 received 2 years after lending $1000. Thus far there should be no riddles, but the next step may require a detailed exposition since what it involves is the stipulation that the payment made at the end of the first year in our 2-year model be put into a savings institution (or financial asset) where *it* draws interest. To begin our explanation, let us use a simple diagram showing what we have done so far (figure 3-1).

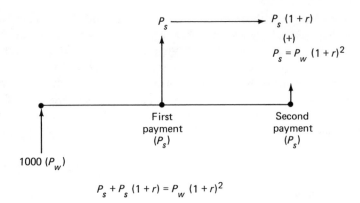

$$P_s + P_s (1 + r) = P_w (1 + r)^2$$

Figure 3-1. Payment Scheme for a 2-Period Model

In the figure, P_w is the amount lent (= \$1000), while P_s is the value of the yearly payment. In line with our earlier discussion, we must have:

$$P_s(1 + r) + P_s = P_w(1 + r)^2$$

or

$$P_s(1 + 0.10) + P_s = 1000(1 + 0.1)^2 = 1210$$

These expressions simply indicate that the payment scheme, handled as in the preceding discussion, provides the same amount of money after 2 years as a perfectly safe bank deposit or bond offering a 10% interest rate. Solving the last equation, we get P_s = \$576. In other words, if you borrow \$1000 that must be repaid in 2 years, you can repay the debt with two end-of-year payments of \$576.

The following discussion will show that, given the conditions of the problem, the first payment must be put into an interest-bearing asset or instrument of one type or another. Let us suppose the first payment were not put into an interest-bearing asset. Then we would have to make the following payments p_s' so that our lender would end up with \$1210 two years after making the loan:

$$P_s' + P_s' = P_w(1 + r)^2 = \$1210$$

or

$$P_s' = 605$$

However, this cannot be satisfactory! If each payment were \$605, the lender could take the first payment, put it in a bank where it would draw interest at 10%, and at the end of the second year the lender would have 605(1 + 0.10) + 605 = \$1270. But \$1270 at the end of 2 years corresponds to an interest rate of 12.7%, which we get by solving $1000(1 + r)^2$ = \$1270 for r. And since, on a single market with perfect information, we can have only one interest rate, the contradiction that we have arrived at indicates that the scheme involving two end-of-year payments that are equivalent to $1210 requires that the first payment draw interest over 1 year.

In other words, two end-of-year payments of $576 in return for a loan of $1000 implies an interest rate of 10%, which is the interest rate with which we began our labors. The reader should now, before continuing to the next paragraph, go through the preceding example using a 3-period situation while keeping everything else the same. Draw the type of diagram in figure 1–3, putting in all the symbols; after calculating P_s, verify that a payment scheme of the sort we have just discussed actually provides the lender with $1000(1 + 0.10)^3$ dollars 3 years after he makes a loan of $1000.

Now let us approach this problem from another direction. Suppose you have a *stock* of money, and you want to turn it into a *flow* of income over a given number of years. With a given rate of interest, you can buy an annuity from, for example, an insurance company, which, in return for a given sum of money, will provide you with a stream of payments at specified intervals over a number of years. For instance, if the interest rate is 10%, a $1000, two-year annuity will yield payments of $576 at the end of each year for 2 years. You can check this result by examining an annuity table.

Now we take up the matter of interest charges along with depreciation and amortization. Referring back to the preceding example, if you borrow $1000 and the rate of interest is 10%, then your interest charges are $100 per year. Thus in terms of the example, of the $576 in total yearly charges, $100 are interest charges, and the rest, $476, are amortization or depreciation charges. To clarify the last concept, let us note that if we have a physical asset, the usual practice is to put money aside every year to replace the asset at the end of a given period. For instance, if we have a machine costing $1000 that lasts 2 years and the interest rate is 10%, if we make yearly depreciation payments (or specify depreciation charges) of P_d and if at the end of the first year the payment is put into a bank where it can draw interest, then if we are to replace the machine at the end of 2 years, we must have:

$$P_d(1 + r) + P_d = \$1000$$

or
$$P_d(1 + 0.1) + P_d = \$1000$$

thus
$$P_d = \$476$$

Note how this result agrees with the previous example in which with $576 in total charges and $100 in interest charges, we arbitrarily called the difference between them ($476) the *depreciation* (or *amortization*) charge. Note that when we speak of a machine, we generally use the expression "depreciation charges";

while with a loan these charges are called "amortization charges" (we speak of "amortizing" a loan).

Finally, before we delve into a little algebra, the preceding discussion can be extended slightly. If we have $1000 and the interest rate is 10%, and if with this money we buy a physical asset such as a machine that is used to produce some product, and if the life of the machine is 2 years—or, as we say, the machine is "fully depreciated" over 2 years—the *capital cost* or *price of capital services* of the machine is equal to $576, or the depreciation cost plus the interest charges. Also, in this example, the *investment cost* is $1000. It would prove useful to spend a few minutes reviewing this terminology because the investment cost and the capital cost are not the same thing, although some people make the mistake of using the terms interchangeably.

Generalizing the preceding discussion, if we take the matter of depreciation or amortization charges first, use the same notation as before, and assume that the asset is depreciated or amortized over T periods, then it follows that we must have:

$$P_d(1+r)^{T-1} + P_d(1+r)^{T-2} + \cdots + P_d = P_m \qquad (3.5)$$

The reader should interpret this expression in the light of the previous numerical example; and then take $T = 3$ and $r = 0.1$, and write out (3.5), keeping in mind our previous discussion. We can now multiply both sides of (3.5) by $(1 + r)$, which yields:

$$P_d(1+r)^{T} + P_d(1+r)^{T-1} + \cdots + P_d(1+r) = P_m(1+r) \qquad (3.6)$$

Subtracting (3.5) from (3.6) gives:

$$P_d(1+r)^{T} - P_d = P_m(1+r) - P_m = rP_m$$

or
$$P_d = \frac{rP_m}{(1+r)^T - 1} \qquad (3.7)$$

The reader should now observe what happens if we take the case of a physical asset with an infinite life: here $T \to \infty$, and on the basis of (3.7), $P_d = 0$. And this is as it should be: If a machine, house, or some other physical asset were never to wear out, then there would not be any depreciation charges. Similarly, if a borrower were to have "forever" to repay a loan, then his amortization charges would be zero. We can now take equation (3.7) and set $T = 2$, $r = 0.10$, and $P_m = 1000$. As in our previous example, we get:

$$P_d = \frac{0.10 \times 1000}{(1.1)^2 - 1} = \frac{100}{0.21} = \$476$$

As explained earlier, our interest cost is rP_m. The total charge, which we have defined as the price of capital services, then becomes:

$$P_s = \frac{rP_m}{(1+r)^T - 1} + \frac{rP_m}{1} = \frac{rP_m(1+r)^T}{(1+r)^T - 1} \qquad (3.8)$$

Again, if $T \to \infty$, then $P_s = rP_m$: For an asset with infinite life, the only charges are interest charges. Then taking the figures from the preceding example, we get:

$$P_s = \frac{0.10 \times 1000\,(1+0.1)^2}{(1+0.1)^2 - 1} = \$576$$

The reader should also appreciate that these results hold regardless of whether we borrow or use our own money to buy a physical asset. In the latter case these charges are called *opportunity costs* since they signify what we could realize if we chose to lend out this money. Finally, let us derive the price of capital services using the convention that in a perfect market, all assets should have the same return. In other words, if we have P_m, which is loaned out for T periods at interest rate r, it should prove the same return as a stream of payment, each amounting to P_s. Thus:

$$P_s(1+r)^{T-1} + P_s(1+r)^{T-2} + \cdots + P_s(1+r) + P_s = P_m(1+r)^T \qquad (3.9)$$

Multiplying (3.9) by $(1+r)$ gives:

$$P_s(1+r)^T + P_s(1+r)^{T-1} + \cdots + P_s(1+r) = P_m(1+r)^{T+1} \qquad (3.10)$$

Subtracting (3.9) from (3.10) gives:

$$P_s(1+r)^T - P_s = P_m(1+r)^T(1+r) - P_m(1+r)^T$$

or
$$P_s = \frac{rP_m(1+r)^T}{(1+r)^T - 1} \qquad (3.11)$$

We can now look at some actual costs for alumina- and aluminum-producing facilities. Table 3–4 shows estimated costs for alumina refineries in North America and the Caribbean with a yearly capacity of 600,000 tonnes per year. These data are for 1969, and with the exception of production costs, are averages. The total investment costs for these installations are \$87 and \$91 million, respectively, which can also be expressed as \$145 per tonne of annual capacity for North America and \$152 per tonne of annual capacity for the Caribbean. These figures

Table 3–4
Production Costs Per Tonne of Alumina, North America, and the Caribbean (1969)
(*All Figures in Dollars*)

	North America	Caribbean
Bauxite (2.5 tonnes/tonne of aluminum)	17.5	7.50
Caustic soda, fuel, other material	11.0	11.25
Maintenance; miscellaneous supplies	2.5	2.50
Labor costs	5.25	4.00
Capital cost/tonne of alumina[a] (depreciation + interest)	17.12	17.85
Miscellaneous overheads	1.00	1.00
Production cost per tonne of alumina	54.7	44.10

[a]Calculated from investment cost/tonne.

Table 3–5
Investment and Operating Costs for Bauxite, Alumina, and Primary Aluminum Capacity (1977)
(*Dollars Per Short Tons of Capacity*)

	Short Tons Required for 1 S-T of Aluminum	Investment Cost Per Ton of Product	Average Operating Cost Per Ton of Product	Investment Cost Per Ton of Aluminum	Average Operating Cost Per Ton of Aluminum
Bauxite	4.5	60	20	270	90.0
Alumina	2.0	500	80	1000	160.0
Primary aluminum	1.0	1800	650	1800	650.0

cover buildings, equipment, and all the normal preproduction activities needed to provide a functioning production unit.

The calculation for the capital cost (that is, the annual price of capital services) for North America, assuming an amortization period of 20 years and an interest rate of 10%, is:

$$P_s = \frac{rP_m(1+r)^T}{(1+r)^T - 1} = \frac{0.10 \times 145(1+0.1)^{20}}{(1+0.1)^{20} - 1} = \$17.12 \text{ per year}$$

Note that here P_m is the investment cost per tonne. Of this capital cost, the interest cost is $rP_m = 0.10 \times 145 = 14.5$. The remainder is the depreciation cost.

Before continuing with data of these type, let us look at some average investment and operating costs for the year 1977. Table 3–5 gives figures for new aluminum-producing capacity.

The reason the investment cost here is so much larger than in the previous example is that there was a huge escalation in capital, labor, and energy costs between 1969 and 1977; the figures in table 3–5 are worldwide averages; and a specific installation size was considered in the previous example. It should also be noted that approximately 2 kW of electric power are needed to produce 1 S-T of primary aluminum per year. In 1977 the investment cost of new electrical generating capacity was from $400 to $600 per kilowatt; and thus each annual ton of *new* aluminum-producing capacity required a further investment of $800 to $1200 per ton. In contrast, the U.S. Bureau of Mines has estimated that the investment cost of aluminum scrap-melting facilities is, on the average, $150 per ton of average melting capacity.

As already implied, the size of an installation is important in determining, for example, the capital cost and the unit-production costs. Data here are not easy to come by, but figures from the United States in the early 1960s provide a definitive clue as to the presence of scale economies in alumina production (table 3–6).

Table 3–6
Some Average Costs for Alumina Plants of Different Sizes in the United States in the Early 1960s
(Dollars/Tonne)

Plant Size (Tonnes Per Year)	60,000	100,000	330,000
Investment cost (dollars/tonne)	217	185	130
Wage cost (tonne of output)	6	5	3
Electric power (Dollars/Tonne)	300	265	200

Note: "Capacity" is equivalent to "size."

Table 3–7
Some Total Unit Costs of Aluminum Production in the United States, Europe, the Middle East, and Australia (1971–1972)
(In Cents/Pound)

	United States	Europe	Middle East	Australia
Alumina	7.50	7.90	7.00	6.20
Power consumption	3.60	4.80	2.00	4.00
Other raw materials	1.50	1.50	2.00	1.50
Labor and overhead	2.75	2.75	2.75	2.75
Depreciation + interest (capital cost)[a]	6.36	5.40	6.94	6.94
Total unit cost (¢/lb)	21.71	22.35	20.69	21.39
Investment cost (per tonne of capacity, in dollars)	1100	950	1200[b]	1200[b]

[a]Assumes 20-year amortization period and 8% interest rate.
[b]Infrastructure included.

At the present time alumina plants tend to be much larger than those observed in this example: Capacities of a million tons per year are becoming common; but as far as is known, scale economies are still important. A similar phenomenon exists for aluminum smelters; and apparently capital costs drop rapidly when we move from small to very large installations. What we will consider now, however, are aluminum smelter costs as they exist in various parts of the world, as shown in table 3-7, where the botton line gives the average unit cost for 1971-1972 at the point of production.

Entries that might be added to this table to give a more rounded picture of costs include the freight on ingot, which in the United States and Europe is mainly an internal cost, and the ocean and internal market freight charges for Australia and some other suppliers, as well as the tariff on aluminum metal. These costs will be discussed in the next chapter on the international trade in bauxite and aluminum.

Appendix 3A*
A Note on Economic Profit

The problem that this appendix will attack is the generalization of some numerical examples constructed earlier in the book. To begin, let us take a situation in which money is borrowed to purchase an asset such as a machine and in which the sum involved is B. Take r as the rate of interest, and assume that there are variable costs E connected with keeping the machine in operation such as those associated with labor costs, raw-material input, power, and so on. To keep this example simple, let us also assume that the asset depreciates completely, that is, falls apart completely at the end of the first period or the end of the first year. Then assuming all payments associated with this asset to be made at the end of the period and with the asset earning R in revenue during the period, the discounted economic profit on this asset during its 1-period lifetime is:

$$V = \frac{R - B - rB - E}{1 + r}$$

In this expression B, which is the cost of the asset, also represents the amortization of the debt that was incurred in borrowing to purchase the asset. Similarly, rB is the interest charge on B. Rearranging the expression, we get:

$$V = \frac{R - E}{1 + r} - B$$

The term $R - E$ is called the *quasi-rent*. Then if $R - E/(1 + r)$ is greater than B, then V is greater than zero. This in turn indicates a *positive profit* on the asset, that is, the yield of the asset is greater than the yield on a financial asset in a perfect market (such as a bond or bank account), and thus the asset should be purchased. If, for instance, we took $r = 10\%$, then if $R - E/(1 + r) > B$, the yield on this physical asset is greater than 10%.

In looking at, for example, costs and benefits in the preceding expression, we see that the interest charges (rB) have disappeared, and in this particular form of the equation, it is unnecessary to designate B as the amortization charge, and sufficient to identify it as the cost of the asset. Of course, an accountant interested in business expenses but not enjoying a grounding in economic theory might find all this a bit peculiar; but remember that what we want to do is to compare the yield of a capital asset with that of a "safe" financial asset, and assuming that the first equation in this appendix, which contains amortization and interest charges, is logically sound, the condition derived from it cannot be disputed.

Now for the important point: Since the quasi-rent, which is obviously some kind of net return, does not involve either the interest or amortization charges, it does not matter whether the money to buy the asset was borrowed or originated internally. Thus in the example in this appendix, the source of the funds for purchasing the asset is irrelevant, and someone trying to work out whether the asset should be acquired or not should not be concerned with this matter. It is sufficient to know that with a given, or chosen, discount rate r, economic profit is positive or zero for those assets actually acquired.

4 International Trade

The production of bauxite depends, quite naturally, on the location of the ore. The decision as to how much will be produced during a given period depends to a certain extent on production costs and the price that the commodity fetches on the market. What should be noted here is the expression "to a certain extent" since it seems likely that the situation with bauxite is similar to that of copper ore for many of its LDC producers: For both economic and political reasons, the price of the ore would have to fall by a very great amount before these countries would willingly cut their output by a sizable increment since to do so would jeopardize their principal source of foreign exchange.

At the same time there are a number of energy-poor, high labor cost, and very pollution-conscious countries with considerable alumina- and aluminum-production capacity. The desire to produce aluminum close to the markets on which it will be sold explains some of this since not only are the transport costs of aluminum metal minimized but it is also possible to maintain a high degree of contact with the changing demands of aluminum consumers. (This last factor is even more important for the production and sale of aluminum semifabricates.) In addition, the market power of aluminum producers vis-à-vis their suppliers of alumina and, especially, bauxite, has generally been greater than that of bauxite and alumina producers relative to the buyers of these products, which means that it has been the aluminum producers who make the decisions as to where processing capacity will be installed. As indicated in the previous chapter, this situation may be changing at the present time, although given the extensive long-run possibilities for substituting out of bauxite and into other aluminum-bearing clays and ores that are to be found in the major industrial countries, a major shift in aluminum production toward the bauxite-producing areas that will not pose a threat to the profitability of all aluminum-producing facilities will require the cooperation of present producers.

The Trade in Bauxite, Alumina, and Aluminum

Table 4–1 shows the imports (and exports) of bauxite by the most important importing (and exporting) countries. During the early 1970s four countries answered for 77% of the world's bauxite exports (Australia, Jamaica, Surinam, and Guyana). Similarly, only a few countries answered for 84% of the world's bauxite imports: the United States, Japan, Canada, and West Germany. Some ex-

Table 4–1
Exports and Imports of Bauxite 1976, and Linear Projections for 1985
(*Thousands of Tons*)

	Exports		Imports	
	1976	1985 (Projected)	1976	1985 (Projected)
Developed countries	9339	16650	27622	45250
United States	70	50	13145	22250
Canada	0	0	1230	5000
Japan	1	0	4277	5300
Australia	6859	14700	0	0
EC	70	100	8720	9400
Yugoslavia	1024	300	41	0
Other developed	1315	1500	209	3300
LDCs	22679	45600	267	8000
IBA members	21944	38400	0	0
Other LDCs	735	7200	267	8000
Centrally planned economies	633	900	5014	9900
World total	32651	63150	32903	63150

Source: Unctad and World Bank statistics. F.E. Banks, *Scarcity, Energy, and Economic Progress* (Lexington, Mass.: Lexington Books, D.C. Heath, 1977).

plations for this situation have beeng given previously, but, everything considered, some question must be raised as to how much longer countries like Japan and West Germany, can continue to play a leading role in the processing of bauxite. It is also interesting to note that many of the exporting countries route the greater part of their shipments to one or two countries. For example, almost all Jamaica's output usually goes to the United States while more than 60% of Guyana's is directed to Canada.

The same pattern can be observed on the importing side. Of both Japan and West Germany's bauxite supply, 60% comes from Australia. The main explanation for this pattern has to do with the role played by transnational corporations in the bauxite-alumina-aluminum market in the sense that they are moving various products between facilities in their ownership or under their control. From the standpoint of global resource allocation, this situation may not be ideal since it almost certaintly involves much more haulage than would take place if shipping patterns were determined merely by minimizing traffic miles, given the constraints provided by shipping costs and the quantity available at various sources and required elsewhere by various processors. The global pattern of bauxite exports and imports are shown in table 4–1.

Shifting from bauxite trade to alumina trade, new countries appear due to the fact that there are important changes on the consumption side of the market.

Energy-rich countries such as Canada and Norway show up as importers (and processors) of alumina; but we can still see the kind of historical-cum-business relationship mentioned earlier in the case of Surinam, a former Dutch colony and a bauxite producer of the first rank, which exports a great deal of alumina to Holland for processing into aluminum.

The position of the United States is also noteworthy since it is both an importer and exporter of alumina. The reason for this can be traced to the structure of ownership of that particular industry with alumina and aluminum capacity on both sides of the border controlled by the same firms. Similarly, given the size of the United States, it makes sense for some alumina producers to send their products to smelters in Canada instead of to other parts of the same country. The matrix in table 4-2 shows the pattern of alumina imports and exports in 1971.

Some general remarks are probably in order at this stage. In particular, some attempt should be made to tie the actual pattern of trade in the bauxite-alumina-aluminum industry to conventional international trade theory. If we attempt to use the theory of comparative advantage to explain trade flows, in particular the Hecksher-Ohlin doctrine, we are almost certain to come up with the wrong answers. In the present situation we are not only contending with a highly oligopolistic industry but one in which economies of scale are extremely important This by itself would tend to invalidate most of the so called wisdom imparted by the better-known models of international trade; but a virtually insuperable barrier to the systematic employment of conventional doctrine is raised by the well-known fact that there are important variations in the way certain items are

Table 4-2
Export and Import of Alumina (1971)
(*Millions of Tons*)

Exports From	Imports to					
	Canada	Japan	Norway	West Germany Great Britain Holland	U.S.	Total
Australia	0.26	0.50	–	–	1.00	1.76
France	–	–	–	–	0.10	0.10
Guinea	–	–	0.17	0.13	–	0.30
Guyana	0.06	–	0.15	–	–	0.21
Jamaica	0.34	–	0.39	0.20	0.45	1.38
Surinam	–	–	0.17	0.39	0.40	0.96
United States	0.28	–	0.19	–	–	0.47
Column total	0.94	0.50	1.07	0.72	1.95	5.18

Source: Unctad and World Banks statistics. F.E. Banks, *Scarcity, Energy, and Economic Progress* (Lexington, Mass.: Lexington Books, D.C. Heath, 1977).

produced among different countries due to the uneven rate at which technology is diffused throughout the industry *and* the world. (Conventional theory, in case the reader has forgotten, limits itself to a make believe world of perfect competition, no scale economies, and identical technologies for all producers.)

In addition, transportation arranagements must be given explicit treatment. Metals and ores are shipped in liners, charter vessels, and industry-owned ships. Of these, ships owned by companies have become increasingly important since World War II for the obvious reason that their schedules are more flexible than liners and their rates more predictable than charter vessels. (In addition, they can be chartered out to other industries in periods when their services are not required by their owners.) Some firms have their own shipping lines such as the Alcoa Steamship Company, owned by Alcoa Aluminum; and Saguenay Shipping Limited, which is a subsidy of Aluminum Limited. Where this practice prevails, shipping rates are established by intracompany accounting procedures, and these are highly influenced, if not determined entirely, by tax regulations of one type or another. It also seems that transport costs play a minimal role in determining shipping routes for both bauxite and aluminum metals. In fact, there have been claims that the present pattern of trade deviates so much from the optimal that the international aggregate of transport costs is between 10 and 25% higher than it could be.

Political considerations also disturb many of the very neat results passed down to us by some of the pure theorists of international trade. The very presence of the (British) Commonwealth, French, and Socialist Blocs is extremely important in determining who gets what of the mineral and metal output of the countries affiliated with these groupings. France, for example, definitely has a preference for minerals originating in its former colonies or protectorates. Moreover, these considerations and others have led to the widespread introduction of long-term contractual arrangements and a structure of ownership that definitely diminishes the speed of reaction of trading partners to such things as price changes.

We will next examine trade flows involving aluminum metal. These flows show that the large alumina importing countries have become major exporters of metal and with certain exceptions, the large exporters of alumina re importing aluminum. Table 4-3 gives exports and imports for 1976, as well as projections for 1985.

This portion of our discussion can be closed by pointing out that in 1972 the OECD countries imported bauxite for a value of 372 million U.S. dollars, and during the same year they exported aluminum metal for a value of 1062 million U.S. dollars. It might therefore be true that there are at least some bauxite-rich countries that should make a detailed examination of the advantages, and disadvantages, of transforming this ore to aluminum metal rather than exporting it in unprocessed form.

Table 4--3
Exports and Imports of Primary Aluminum, 1976, and Linear Projections for 1985
(*Thousands of Tons*)

	Exports		Imports	
	1976	*1985 (Projected)*	*1976*	*1985 (Projected)*
Developed countries	2670	4046	2698	5572
United States	139	150	517	1374
Canada	508	1135	22	20
Japan	70	50	430	1596
Australia	64	220	1	3
EC	909	900	1521	2249
Yugoslavia	72	257	32	30
Other developed	908	1334	175	300
LDCs	347	1446	96	150
IBA members	179	465	0	0
Other LDCs	168	981	96	150
Centrally planned economies	696	773	188	600
World total	3713	6265	2982	6322

Source: Unctad, World Bank, and OECD documents.

Commodity Politics and the Trade-in Bauxite

This section also deals with the trade in bauxite but in the context of international-commodity politics. The starting point for our discussion is OPEC. This organization, which was established for the purpose of protesting the loss of oil-producing countries of a few cents of revenue per barrel of petroleum, was able to achieve a remarkable increase in both its economic and political power *without* the permission of the oil companies *or* the industrial world. In 1973-1974 OPEC raised the net selling price of oil from almost $2 to slightly less than $12 per barrel. This increase gave the countries belonging to OPEC an income from oil that was well in excess of $100 billion, in addition to an enormous political leverage in the capitals of their main clients. Considering that the average unit cost of lifting a barrel of oil is under $1 in the OPEC countries, this cartel has mobilized a sizable surplus for itself out of which such things as capital investments, consumer durables, and military hardware can be financed.

Other countries producing industrial raw materials, after observing all this, have made no secret of their desire to acquire the same status. Some of the producers of bauxite have formalized their intention to take full responsibility for the raw materials located within their national boundaries as well as bringing about a large increase in their export revenues by forming a producer association that they call IBA (The International Bauxite Association). Dating from 1974, its original members were Jamaica, Surinam, Guinea, Guyana, Australia, Sierra Leone, and Yugoslavia. Later on the Dominican Republic, Ghana, Haiti, and Indonesia were added to this roster.

IBA members produce about 60 million tons of bauxite a year. Before they rearranged the taxes being paid by the bauxite-producing firms operating in IBA countries, the value of their export income (FOB) was on the order of $0.6 billion, or approximately one thirtieth of the income of the OPEC oil producers prior to the 1973–1974 oil price rises. After the imposition of new taxes by various IBA countries, revenues increased by almost $400 million—which is not much when compared to the increase in OPEC revenues mentioned earlier. It also appears that the initial income increase enjoyed by Jamaica, the country taking the lead in increasing taxes, was mostly absorbed by Jamaican payments for oil imports at the new high prices. The mechanics of the Jamaican tax increases are fairly straightforward and amount to the following. In addition to a standard royalty consisting of 50 Jamaican cents per dry long ton (LDT) of all bauxite mined, they imposed a production levy consisting of 7.5% of the average realized price (in U.S. dollars per short ton) of primary aluminum ingots divided by 4.3. (In addition, all firms are required to produce a minimum tonnage of bauxite, which was initially specified as 90% of the annual production of each company for 1974.) Roughly, what this means is that $2 will be added to the cost of the 345 pounds of bauxite needed to produce the 80 pounds of aluminum used in a typical U.S. automobile in 1974–1975, and presumably this would be passed on to the purchaser of the vehicle.

In trying to grasp the rationale behind these tax increases, certain things seem evident. To begin, there was never a question of an embargo on bauxite. Not only do most bauxite-consuming countries possess large stocks of this commodity, or of aluminum, but as pointed out earlier, the stock of secondary aluminum that could be made available in an emergency is huge. Equally important, a complete bauxite embargo would immediately increase bauxite prices to a degree that the introduction of processes employing other aluminum-bearing clays and ores would be accelerated, which is something that bauxite-producing countries intend to avoid at all costs.

The levying of taxes that raised bauxite prices by an average of $10 a ton in 1974 and about $25 a ton in January 1976 thus struck the bauxite-producing countries as a relatively painless manner of augmenting their income from the sale of this asset. But there were also in some cases changes in the structure of ownership of the bauxite industry. Without bothering to secure IBA approval, Jamaica nationalized *all* those properties belonging to aluminum companies that were not being used, either directly or indirectly, in the production and/or transportation of bauxite or alumina; and they also nationalized 51% of those bauxite-alumina facilities that were in full production. The producing companies were "compensated" with low-interest notes whose honoring depended on their selling enough aluminum to provide Jamaica with a large tax income. In 1975–1976 the "extra" tax income of Jamaica was $130 million. Similarly, Surinam and Guinea realized $60 million each, Guyana $30 million, and the Dominican Republic and Haiti $10 million and $6 million, respectively.

In addition, all producing countries did not impose the same levies. Furthermore, it has been said that the Jamaican levy was adjusted downward when they realized that other countries were applying lower rates and were prepared to expand their output by a very large increment. (Moreover, countries such as Brazil, with a very high bauxite-producing potential, announced that substantial additions to capacity would soon be commenced.) Guinea's buaxite tax, for example, averaged about $6.50 per ton, but since Guinea's bauxite is a higher grade than Jamaica's and is farther away from North America (where Jamaica sells much of its bauxite), the effective tax rate relative to Jamaica was lower. Surinam's announced tax arrangement consisted of a 6% levy while, as mentioned earlier, Australia has no levies, although the state of Queensland has put a tax of between $0.5 and $1.50 per ton on extracted ore.

The issue now is just how far the bauxite-producing countries can go with their attempts to increase their revenues via taxes and royalties. In the short run alumina plants are specialized to a few principal varieties of bauxite. Alumina-producing companies are thus not prone to antagonize their bauxite suppliers since the profitable functioning of hundreds of millions of dollars of processing capacity depends on an uninterrupted flow of the right kind of feedstock. This situation has greatly favored some of the Caribbean countries, particularly Jamaica. Other countries, however, might find it expedient to go slow on their taxing or nationalization plans. There are a number of important people in Australia who insist on raising the already high level of mineral production, and as things stand, a program of this type will require a large injection of foreign capital. But were the most eligible providers of this capital to get the idea that business risk was capable of being arbitrarily increased from time to time by major tax boosts, they might prefer to concentrate their attention on some other part of the world or even to forget about bauxite and promote other aluminum-bearing materials. There is enough alunite and anorthosite in the United States to supply the entire world with aluminum for the next hundred years or so at present use rates, and much of this appears to be in the vicinity of large deposits of inexpensive coal. Were it possible to move the processing of these nonbauxite assets past the pilot stage, which is the object of a number of major research projects around the world, then the owners of bauxite might find their market power fatally diminished. These owners know this, and an official of the United States Bureau of Mines has said that one of the reasons that the search for alternatives to bauxite must continue is to keep bauxite prices from exploding.

We can conclude this phase of our exposition by making a final comment on Australia's place on the bauxite-producing scene. Although Australia is the largest producer of bauxite in the world, that country has pursued a price policy that is markedly different from most of the other producers. In fact Australia, which was one of the original members of the IBA, announced in January 1979 that it was no longer prepared to follow IBA operational outlines because of its rather special geographic situation. (Brazil has also declared its independence from the IBA, and present intentions are to ship 5 million tons of bauxite yearly by the late 1980s.) As mentioned earlier, bauxite is only a portion of the Australian mineral picture, and caution must be exercised by the government of that country lest foreigners begin to think of Australians as being unfriendly to foreign capital. There is also the matter of what is the optimum strategy for Australia on a market whose producer side is dominated by a fairly aggressive cartel. This topic has been examined in some detail by Pindyck (1976, 1977), and his findings can be summed up as follows.

Australia can gain marginally from *not* following cartel-pricing policies if its long run elasticity of supply is the same as other producers, and gain significantly if its supply is more elastic—that is, if it is able to increase its output by fairly large amounts without incurring large increases in marginal costs. The common sense of this observation clearly involves no more than Australia's raising its production and to a certain extent its price as the others raise their price. Thus Australia eats into the market share of the others. Or, conversely, the countries moving their price up the most rapidly suffer a declining market share, as seems to be the case with the Caribbean countries today.

Multinational Companies and the Structure of the Aluminum Industry

Earlier this chapter pointed out that there seemed to be a serious departure from an "ideal" trading pattern in the bauxite-alumina-aluminum industry due to the presence of such things as political blocs and influence and multinational companies (MNCs). Since this is a textbook in economics and not political science, only the latter issue will be taken up although, as the reader might guess, there is considerable interrelation between them.

We can start out by noting that the era of the MNC began in the 1960s. These organizations had been powerful since the beginning of this century, but it was during that period when world trade and international economic development seemed to be reaching a kind of crescendo that a few large firms operating in international markets were able to achieve revenues in excess of the budgets of some fairly important countries and the gross national products of some lesser domains. The key to all this was coming into possession of the right kind of currencies (principally dollars), and using these financial assets to acquire other assets such as capital equipment, structures, and properties outside their home countries. Canada provides a convenient example of this process. By 1966 almost one half of the Canadian exports of $5.3 billion were produced by foreign-owned corporations, most of whom had their head offices in the United States.

Naturally arrangements of this type provoke a certain amount of bad feeling. The issue usually cited is loss of national sovereignty; but the opinion here is that there are very few, if any, countries that over the past decade have lost any sovereignty due to MNCs' plotting with foreign countries or domestic political groupings. Furthermore, since the day of gunboat diplomacy seems to have passed us by, I would argue that there are no rational governments anywhere who are prepared to place the property rights of MNCs before domestic political expediency.

Of course, as indicated earlier, there is a strong possibility that countries like Australia might have been prevented from carrying out certain desirable economic programs because foreigners have too much to say about where Australia's mineral output should be processed. But as far as I am concerned, if these outsiders had no weight whatsoever in this matter and if every cubic centimeter of ore extracted in Australia were processed domestically, the gains to the Australian economy would be trivial in comparison to those that would be realized if the primary and secondary education system of that country were revamped in such a way as to produce people capable of playing a more productive role in the community. (And this observation is equally true for most countries in the industrial world.) This may seem like an irrelevant point to be making in a book of this nature, but it seems to be true in Australia that MNCs not only supply capital but are also in possession of certain specialized knowledge that is apparently not easily obtainable locally.

Next we can survey the structure of the aluminum industry, which in practice means saying something about very large and economically powerful MNCs. There are more than 70 firms producing primary aluminum, but in 1973 six of these accounted for 60% of the production of the major industrial countries. This percentage may sound very high, but 15 years earlier the figure came to 80%.

The explanation for this state of affairs turns on the "indivisibilities" argument of microeconomic theory. A world of backyard smelters and semifabricators might improve competition (although personally I have my doubts as to whether it would), but it would hardly reduce the price of aluminum metal *or* the price of products containing aluminum metal. Aluminum-processing facilities require huge amounts of capital, and most evidence indicates that they display diminishing unit costs as installation size increases. Unit costs are also affected favorably if processing facilities are assured a steady flow of feedstock, because both capital and labor can be more economically utilized. This in fact is one of the reasons why some of the larger aluminum firms have been so aggressive in acquiring bauxite- and alumina-producing properties. It also seems to be true that large firms, with their production problems firmly under control, can obtain cheaper credit than smaller producers.

Before continuing this phase of our discussion, note that the decrease in the market share of the six major aluminum producers has been accompanied by an increase in the growth of state-owned companies and companies in which the state has a decision-making role. This phenomenon has occurred in such diverse countries as Germany, Austria, Italy, Spain, Norway, South Africa, Bahrain, India, and Iran. Strong integration tendencies have also been noted on the part of semifabricators, although this phenomenon is not new: Giants like Kaiser and Reynolds entered the primary aluminum industry as semifabricators. And we should not overlook the growing importance of Japanese aluminum producers. The four major Japanese aluminum companies (Sumimoto Chemical Corporation; Nippon Light Metal—with Alcan holdinga 50% interest; Mitsubishi Chemical; and Showa Denko) have shown a remarkable rate of expansion, although with the exception of Nippon Light Metal, aluminum contributes less than a third to their sales.

Continuing with our main topic, it can be argued that the trend toward both horizontal and vertical integration in this industry accelerated in pace with the liberalization of trade. The aluminum market is such that a firm restricted to domestic operations suffers a clear competitive disadvantage when confronted by organizations with an international perspective. Once a few firms had gained access to the global market for inputs and outputs, many others were literally forced to augment their international production and distribution networks. Thus it was the entry of North American MNCs into the European market, which was facilitated by the free-trade euphoria of the postwar world, that compelled firms like Pechiney to take an interest in semifabrication and convinced Vereinigte Aluminum Werke (VAW) that it was a mistake to limit its activities to the German market. Subsequently, that company established distribution centers in neighboring countries and even became involved in a number of bauxite projects outside Europe.

In examining what all this has meant for the aluminum industry, it could be suggested that capacity has shown a tendency to expand in concert with demand to an extent that may not have been possible had decision making been less decentralized. This expansion occurred because of the need to coordinate movements between geographically diverse sources of various inputs while at the same time handling quantities large enough to make advantage of the considerable economies of scale realizable in both production and transport. It may be so, however, that the slowdown in international economic activity will uncover some weaknesses in the present ownership arrangements. Until recently, new capacity was being added to the industry at a fairly high rate, and aluminum metal was being sold in such a way that its price was often so low as to pose a serious threat to the profitability of older installations. A low price, of course, favors the consumer in the short run, but perhaps not in the long run. As will be made clear later, one of the principal effects of low prices and low profitabilities is a diminished rate of investment in new facilities, which can often mean severe shortages and much higher prices in the future. At the present time, for example, there is a very strong opinion in some industry circles that the depressed prices, low profitability, and excess capacity that characterized the industry from 1975 to 1978 will mean a shortage of capacity in the early 1980s.

We can conclude this section by saying something about the six largest aluminum companies. It was once pointed out to me by Professor Jacques Royer that a large part of the subject of international economics might some day consist of the study of MNCs; and nowhere is this profound observation more correct than in the field of nonfuel minerals. As for the major aluminum firms, they are:

1. *Aluminum Company of America (ALCOA)*. In 1974 this firm possessed a primary aluminum capacity of 1705 thousand metric tons. Its sales came to $2730 million, and its average annual growth over the period 1969-1974 was 11.7%. Aluminum products and components contributed 95% to its sales, and it is a completely integrated company, operating bauxite mines as well as producing manufactured and semimanufactured products in aluminum and its alloys. The company also has interests in shipping and real estate, and operates in 17 countries. Its principal facilities though are to be found in the United States, Australia, Brazil, Norway, Mexico, Surinam, and the United Kingdom.

2. *Alcan Aluminum Limited (Canada)*. This holding company has interests in 100 firms engaged in bauxite mining, processing and fabricating aluminum, producing electric power, and transporting and selling the group's products. Its sales were $2430 million (U.S. currency) in 1974, and it had 1520 thousand metric tons of primary aluminum capacity. Operations are carried out in more than 30 countries, with the most prominent being Canada, Japan, Norway, the United Kingdom, India, Brazil, Australia, Spain, Sweden, and England.

3. *Reynolds Metals* (U.S.). Reynolds is the second largest producer of aluminum in the United States. In 1973 this corporation possessed a primary producing capacity of 1148 thousand metric tons, which generated a sales of $1995 million. Its average annual growth rate in the period 1969-1974 was 14.5%, which was the highest among the major firms. Reynolds is an integrated company that conducts operations in bauxite mining and alumina and aluminum production in 23 countries, the most important of which are the United States, Canada, Norway, Ghana, the United Kingdom, and Venezuela.

4. *Kaiser Metals* (U.S.). Kaiser differs from the other North American producers in that aluminum contributes only 70% to its total sales. In the United States, Kaiser is one of the largest manufacturers of refractories, and it also has large interests in shipping, real estate, and trading. Its total 1974 sales came to $1730 million, and its aluminum-production capacity in 1973

was 962 thousand metric tons. Its field of operation includes the United States, Germany, Ghana, Bahrain, India, and Australia.

5. *Pechiney Ugine Kuhlmann* (France). This firm had sales of $4572 million in 1974, of which 33% came from its aluminum operations. Its aluminum-producing capacity was 953 thousand metric tons in 1973, and Pechiney is also involved in, among others, copper processing, chemicals, and electro-metallurgy. Pechiney is the largest firm of its kind in Europe and conducts operations in France, Greece, Spain, the United States, the Netherlands, and the Cameroons.

6. *Aluswisse* (Switzerland). Aluswisse is active in chemicals and plastics as well as in mining bauxite, producing alumina and aluminum, generating electric power, and fabricating finished products. Its operates in Switzerland, Germany, Italy, the Netherlands, Austria, Norway, Iceland, the United States, and South Africa. Its aluminum-producing capacity in 1973 was 479 thousand metric tons, and total sales were $1679 million. Of this revenue, 80% was associated with the production of aluminum.

Some Political Economy of the Trade in Resources

This chapter will conclude with a brief comment on export-price bargaining. The following discussion is prompted by an interesting paper by Ben Smith (1977); however, in contrast to the apparent conclusion of Smith's paper, the contention here is that information problems make this category of problem basically insoluble—or "indeterminate," to use the relevant expression from economic theory.

The specific case that Smith examines is the mineral trade relations between Japan and Australia, in which both sides are seeking some kind of bargaining advantage in the typical spirit of modern game theory. But what should be carefully noted is that in the game being portrayed, there is little scope for the kind of "misrepresentation" that is a hallmark of the course in "competitive decision making" at the Harvard Business School. Instead, the feature of this situation is the inability of one or both traders to fully comprehend their *own* strength and/or that of the other trader—not to mention any third parties that may be relevant to the analysis to include those that might put in an appearance in the future—because of the difficulty of predicting and evaluating the course of future events. Many of us have now come to understand, in fact, that it is this impossibility that separates the oversimplified world of theoretical economies from crass reality, and nowhere is this more evident than in the field of resource economics. Of course, it has been suggested that this problem may be less difficult for the soothsayers of Australia than Japan since regardless of what the future holds in store, Australians can be counted on to support almost any course of action that involves an expansion of their personal consumption horizons, regardless of who loses in the process or what it costs the country in the long run.

Among Smith's most important observations is his recognition of the futility of fully exploiting monopoly or near monopoly power if the intention is to prolong the trading relationship. At the time of the oil price increases, Japan considered that it was in its long-term interests to temporarily adjust upward the previously established price at which it was buying Australian iron ore, even though the world demand for iron ore was falling. Obviously a decision of this nature involves a number of deviations from the better known results of elementary economic theory. The basic issue here though is simple risk aversion in which the buyer is attempting to influence the altruism of sellers in hypothetical future situations in which he, the buyer, might be at a disadvantage.

One way traders attempt to guard against future uncertainty is to use long-term contracts. The variables involved here are the quantities of the product that will be traded as well as the time of delivery and particularly the price to be paid for the product and the "maturity" of the contract (or the time period for which the contract is applicable). On the subject of price, Smith has taken a "world price" that has been set external to the actions of the two main participants in his scenario. Given the nature of the minerals markets, I prefer to think in terms of a basic suppliers' price determined by the costs of production. Adding transport costs to this gives the basic price at the buyers' doorstep; and to this is usually added some kind of mark-up that is negotiable. On the basis of my own limited knowledge of these matters, I feel that few buyers faced with a small number of sellers, and having any regard for the future and the ups and downs of economic life, would like to acquire the reputation of wanting to acquire more than a minumum amount of their raw-material requirements at a price that did not cover the producers' normal costs, even though at the same time they might be hesitant about granting the seller a profit larger than that necessary to prevent physical and financial resources from deserting the particular industry and seeking a more lucrative outlet.

Similarly, with the exception of LDCs who depend largely on one or two products as a source of foreign exchange, if the price fell below the cost of production and stayed there for awhile, suppliers would tend to not replace their production facilities as they depreciated, which means that production would gradually decline. Conversely, when large profits were being realized, many producers would rush to expand their facilities. This type of reasoning follows from elementary economic theory; and a few costs of the type implicit in the preceding discussion are shown in table 4-4.

Complications enter the picture when we turn to the conditions put into long-term contracts and in particular the determination of the maturity of these contracts. Risk-averse buyers and sellers traditionally prefer contracts with long-running times that specify, within a fairly restricted range, prices and quantities (and the same might be true of some "risk preferers" faced with high search costs). Smith assumes, and correctly I think, that current or incipient market opportunities are readily discernible to all except those businessmen and politicians with a special capacity for misinterpreting reality; and so the key issue is

Table 4–4
Some Prices and Costs for Three Important Minerals

	Bauxite[b] Jamaica-U.S. 1976	Copper[c] Chile-Japan 1972	Tin[d] Malaysia-U.S. 1976
1. Production costs	39	44.0	5331
2. Export price	106[a]	48.5	7309
3. "Economic rent" (2–1)	67	4.5	1978
4. Shipping costs	11	1.0	115
5. Import price (CIF)	–	49.5	7424
6. Production cost in the importing country (total)[e]	909	–	–
7. Local distribution or wholesale costs	–	2.7	585
8. Market price	977	54.0	8373

[a]FOB Kingston.
[b]Monetary unit: U.S. dollars per tonne. Market price: Price of aluminum metal.
[c]Monetary unit: U.S. cents per pound.
[d]Monetary unit: U.S. dollars per tonne (wholesale price).
[e]This item includes the price of imported bauxite. The production cost in the United States, excluding the cost of bauxite and shipping, is $792 per tonne.

actually deciphering the course of events over a half-decade or so. Under the circumstances, in addition to the well-known indeterminacy that would exist if we had a bargaining game involving a known quantity of goods at a given point in time, we have another indeterminacy caused by different rates of time preference on the part of traders as well as different capacities to interpret future events. The way this game is settled (in the sense that contracts with definite maturities *are* signed specifying prices and quantities) depends on the bargaining strengths of the participants, and the basic courses in economic theory have precious little to say about what this amounts to in reality or, for that matter, how it is perceived.

But papers published in learned journals cannot end on this note, and so another approach is warranted. What Smith does is propose a situation where cartel faces cartel instead of firm facing cartel or firm facing firm. The theoretical advantages of cartelization (or monopoly) are obvious from even a superficial inspection of the technical literature or even of real life. This advantage is even more pronounced in the case of exporters or importers since presumably nobody —to include the antitrust departments of most governments—has any really strenuous objections to seeing foreigners treated unfairly; and thus in an attempt to improve their bargaining power, both sides evolve toward a pure cartel, that is, a collection of firms acting under a central decision-making body and effectively becoming a monopoly.

In Smith's work the formation of this monopoly leads to a fall in the volume of trade. In theory, this decline should be expected because, as learned in the first course in economics, if demand curves for factors and goods deviate from the horizontal position that typifies so-called pure competition, we get a lower equilibrium output. However, these results depend on demand varying with price, and in most mineral markets, in the short to medium run, at least, demand is price inelastic, and changes do not take place in the amount traded. Thus we can cease to be concerned with such factors as quantities and focus our attention on price. This is settled by negotiation between the two cartels, and when established determines the profit of each trading "partner."

Smith's resolution of this conflict involves invoking a proposition by Spindler (1974), which takes the following form: In a situation of bilateral monopoly, the trader with the greatest profit has the most to lose in the event of a no-sale situation and thus will always give way. Since this could apply to either trader, the only stable arrangement is one in which both traders have the same amount to lose, which can be interpreted as meaning that an *equilibrium* solution is one in which each trader has the same share of the total profit.

Ideally, this solution is probably as satisfactory as a dozen others floating around in the scientific literaure, but unfortunately it cannot be taken seriously as a description of the way things happen in the real world. Smith recognizes this flaw by pointing out that there may be factors influencing bargaining power that are not considered in the model. This is incontestably true. At the same time Smith contends that since profit maximization is the goal of this bargaining, these other factors are of only minor importance. This is not quite so true since the relevant profit maximization is of the intertemporal variety and involves hypothetical rather than objective cost and demand curves. Thus once again we are back at a situation in which the perceptions of the traders count for at least as much as the objective economic facts of life, and the job of both parties is to convince each other, and themselves, that their vision of the truth is the one that should be taken seriously in determining just who should get what.

This brings us back to square 1: an indeterminate problem! At least in this particular situation, the propositions of basic welfare economics cannot be disputed.

Appendix 4A
Exhaustible Resources

Nothing has yet been said about the ideal rate of extraction of a natural resource; and so before going on to the next two chapters and the topics of supply, demand, and price, this gap will be filled. The problem of optimal extraction was first attacked with the implements of modern mathematical economics by Hotelling (1931), and his work has been updated and expanded by leading economic theorists like Murray Kemp and Robert Solow.

Stripped of unnecessary detail, the problem comes down to the following: If we have a unit of a resource in the ground that can be removed at a certain cost $\bar{M}_t$, which can be interpreted as a marginal cost, and this unit can be sold for a known price p_t, then the decision as to whether the unit should be removed turns on the rate of interest in the following way.

By removing the unit now and selling it at a price p_t, we obtain a *marginal profit* of 1 $[(p_t - \bar{M}_t) = MP_t]$. If we put this marginal profit in a financial institution, our gain over 1 period is $MP_t(1 + r) - MP_t = rMP_t$, where r is the prevailing rate of interest.

If instead we leave the unit in the ground and extract it at the beginning of the next period, we realize a *capital gain* of $(p_{t+1} - \bar{M}_{t+1}) - (p_t - \bar{M}_t) = MP_{t+1} - MP_t$. If $rMP_t > (MP_{t+1} - MP_t)$, then the correct action is to remove the unit; however, if $rMP_t < MP_{t+1} - MP_t$, then the unit of the resource is not extracted. Our *equilibrium* situation, which is the point at which we cease extracting the resource, comes when:

$$MP_{t+1} - MP_t = rMP_t$$

that is

$$\frac{MP_{t+1} - MP_t}{MP_t} = r$$

or

$$\frac{\Delta MP}{MP} = r$$

With zero extraction costs, the last expression reduces to $\Delta p/p = r$, and looking at the last expression, we recognize immediately a difference equation that, allowing time periods to become very small, can be written as a differential equation with the solution:

$$MP = MP(0)e^{rt}$$

or

$$p - \bar{M} = MP(0)e^{rt}$$

that is,
$$p = \bar{M} + MP(0)e^{rt}$$

In these expressions, $MP(0)$ is formally the MP at time zero, although connoisseurs of this type of analysis probably realize that the calculation of $MP(0)$ can be a very sophisticated operation and usually involves finding an $MP(0)$ such that the total quantity of the resource is used up at the end of the last period T. $\bar{M}$, as before, is the marginal cost.

Note that in the case of a resource having a finite supply, $p \neq \bar{M}$, which might seem unusual for students of elementary microeconomics. Instead, marginal cost must be augmented with a *royalty*, which, as explained by Anthony Fisher (1977) and Banks (1974, 1976), compensates for the fact that if a mineral is extracted today, it is not available for extraction tomorrow.

This brings us to the matter of *backstop technologies*. This term was introduced by Nordhaus (1974) in an article having to do with energy resources; but as Robert Pindyck (1976, 1977) makes clear, this usage is quite applicable to the aluminum industry since it is just as plausible to think of a mass introduction of nonbauxite technologies (for example, alunite) as a switchover to, for example, shale oil, tar sands, and nuclear devices.

For expository purposes we shall consider two processes for producing aluminum. One of these uses one unit of bauxite per unit of aluminum, and for simplicity, it will be assumed that bauxite is finite in supply with B units available; but these units can be extracted at zero cost. On the other hand, the second process involves alunite, which is "infinite" in supply but whose extraction requires nondepreciable capital worth K monetary units. Also assume that the rate of interest is r, and each year A units of aluminum are required. (By making this specification, we remove the influence of price on the future demand for aluminum.) Given these data, bauxite will be extracted first, and the switch to alunite, which for the purpose of this exposition has been designated the backstop technology, will take place in B/A years.

Now let us note that, assuming a perfectly competitive market, at a switch point $\hat{T}$, the price of alunite is given by its cost, or $\hat{P} = rK$. (This is the rental cost of nondepreciable capital, and it is assumed that there are no other costs.) Then remembering that we assume no extraction costs for bauxite, this price discounted back to the present t gives us the current cost (or price) of bauxite.

$$p(t) = p(\hat{T})e^{-r(\hat{T}-t)} = rKe^{-r(\hat{T}-t)}$$

Since there are no extraction costs, this result is the "royalty" on bauxite; in other words, since there will be a demand for aluminum even when bauxite is exhausted, sufficient cash must be accumulated to pay for the aluminum-producing equipment. By charging this price and then putting the revenue from bauxite sales in the bank or using it to buy bonds, by the time the existing stock B is exhausted, enough money will be available to purchase equipment that, using alunite as a feedstock, can produce A units of aluminum per year. We also note that

$$\frac{dp(t)}{dB} = rKe^{-r(\hat{T}-t)}$$

In other words, if B decreases (or A increases), $p(t)$ decreases.

5

Demand, Supply, and Price

This chapter examines the demand for aluminum as well as some additional aspects of supply. Price formation on the market for nonfuel minerals will also be introduced and continued in the next chapter. The latter topic should be of particular importance to readers specifically interested in aluminum since the recent decision to introduce an aluminum "contract" on the London Metal Exchange means that the mechanics of pricing this metal should soon be similar to that of a number of others, in particular, copper. At the same time the manner in which these metals are priced differs considerably from the explanations provided in almost all textbooks, and those interested in understanding how actual commodity markets work should pay particular attention to the discussion in the last part of this chapter and the first part of the next.

In the course of discussing the demand for aluminum, some elementary theoretical materials on elasticities will be presented. The issue here is basically simple and can be summed up as follows: In the short run, aluminum's price elasticity of demand—or the response of demand to a change in price—is very low. That is, there is only a relatively small, even insignificant, change in the demand for this metal when there is anything except a very large increase or decrease in price. One ramification of this fact is that, unlike economists, aluminum producers have not been concerned with the possibility that small, random increases in the price of aluminum would result in the substitution of copper or some other metal for aluminum; and the same is true of copper producers. What is happening, in fact, is that most analysts are beginning to regard aluminum as the metal of the future since it is gradually displacing copper in many of its uses despite occasional shifts in the price ratio of these two materials in favor of copper.

Consumption

If we look at the average annual increase in the consumption of some important metals over the period 1951-1971, we see the following:

	Aluminum	*Zinc*	*Copper*	*Lead*	*Tin*
Average annual growth Rate of consumption (%) (1951-1971)	8.8	4.0	3.7	3.0	1.3

Present growth rates for several metals are somewhat below these figures, the trend having been broken by the energy crisis of 1973-1974 and the subsequent deceleration of the world economy. As already indicated, the demand for aluminum is quite insensitive to price, but it is strongly influenced by changes in industrial production and/or national incomes. The reason can be seen immediately when we view the pattern of aluminum consumption: A large share of the output of this metal is used in, for example, building construction and electrical machinery, and the demand for these items is very sensitive to movements in the business cycle. Table 5-1 shows both the level of aluminum demand for the western industrial countries and the levels and trend growth rates for the subgroups Western Europe, the United States, and Japan.

We can now take up some of the details behind the figures in table 5-1. The transportation industry comprises the largest market for aluminum metal, and it is of such great importance for the aluminum industry that a trend movement away from the use of aluminum in automobiles appears to have been reversed. In 1920 the average aluminum content of an automobile in the United States was about 18 kg, which had declined to approximately 5.5 kg

Table 5-1
The Demand for Aluminum by Various Sectors (1973)
(*In Thousands of Tons*)

Sector	Western Industrial Countries	Western Europe	United States	Japan
Transport	3036	1154	1200	344
Average annual growth rate (%)		6.1	5.8	20.5
Mechanical engineering	828	330	345	87
Average annual growth rate (%)		4.6	6.1	16.0
Electrical engineering	1656	454	670	226
Average annual growth rate (%)		6.5	8.7	24.0
Building and construction	3589	700	1590	654
Average annual growth rate (%)		12.9	8.7	40.0
Packaging	1794	454	970	28
Average annual growth rate (%)		7.9	16.1	14.8
Domestic and office equipment	966	330	415	141
Average annual growth rate (%)		5.5	4.1	2.7
Other	1933	700	546	381
Average annual growth rate (%)		6.5	6.5	19.7
Total demand (000 tons)	13802	4122	5736	1861
Average annual growth rate (weighted average)		7.0	7.9	19.5

Source: World Bank documents.

by 1948; but at the present time this figure has recovered to an industry average of almost 45 kg. With vehicle manufactures and their customers beginning to favor smaller and lighter cars, aluminum should find increased application in the construction of engine blocks, bumpers, wheels, and various fixtures and fittings. According to the prestigious research organization *Euroeconomics,* there should be a 4% average annual growth in aluminum consumption per passenger car over the next decade or so; and on the basis of industry reports, it seems likely that over this period, the amount of aluminum that will be found in automobiles manufactured in Detroit will increase to an average of 127 lb per vehicle.

Similar observations pertain to the commercial-vehicle sector. In the western industrial countries commercial vehicles account for a quarter of total vehicle production; while in terms of weight their share is approximately one third. Due to the growing importance of public transport for urban areas and the increasing popularity of long-haul freight operations, prospects for this sector remain good; and since it appears that a more intensive use of aluminum means not only a smaller input of fuel per passenger mile but decreased operational and maintenance costs, the increase in aluminum consumption per vehicle should be at least as great as with passenger automobiles.

In aviation there was a slowdown in the use of aluminum until a decade or so ago. This slowdown was caused by the replacement of aluminum by plastics in various airframe parts and, for both structural and thermodynamic reasons, by stainless steel and titanium in jet engines. But apparently some of the ground lost here is being regained as a result of the introduction of new alloys. The principal factor determining how well things will go for aluminum in the world aviation industry, however, is the yearly increase in the number of airline passengers. This increase showed a tendency to stagnate in 1974-1975, but the introduction of a lower level of air fares might possibly restimulate the demand for air travel in the near future, just as the introduction by many airlines of new equipment should revive the demand for aircraft inputs in the early 1980s.

In building and construction, the main uses of aluminum are in doors, windows and frames, in gutters, and in house fittings and fixtures in general. Aluminum is used in these areas because of its homogeneity (as compared, for example, to wood), its water repellency, its adaptability to precision fitting of parts, and its aesthetic qualities. Some aluminum is used in the main structural members of both buildings and bridges, but given the relative cost and strength of ordinary steel, there is some question as to aluminum's future in this application. The pricnipal competitors of aluminum in this sector are plastics and stainless steel—the latter because of its low maintenance costs and immunity to corrosion. In line with the observations made earlier on the transportation sector, the future of aluminum here is tied to the resumption of a high level of construction activity. At present, though, there is a downturn in building throughout most of the industrial world; and there is serious talk about an irreversible drop in the average annual growth rate of the construction industry.

As seen in table 5-1, the manufacturers of electrical equipment use a great deal of aluminum. It is also true that there are many applications in this sector in which aluminum is in direct competition with copper. The low weight of aluminum relative to copper makes it particularly suitable for high-tension transmission lines, despite the fact that it has only 63% of the electrical conductivity of copper. Where underground uses are concerned, aluminum has met with less success, the principal difficulties arising in the form of insulation problems and excessive losses in transmission. Where this application is concerned, it has been said that most of the opportunities for a large-scale use of copper have been utilized, and future developments will be more serene.

Aluminum has also begun to make important inroads into house wiring and as conductors in electrical equipment. The relative high price of copper during the period 1967-1972 was behind the successes scored by aluminum in this field, but the recent major decreases in the price of copper may alter this development. Aluminum house wiring is widespread in France and important in the United Kingdom, but there have been problems with household fittings and a certain resistance against working with aluminum on the part of electricians in the United States. In electrical machinery and equipment, aluminum should be able to make considerable advances in the manufacture of transformers and busbars. Table 5-2 provides some insight into the physical and mechanical properties of aluminum relative to copper.

A field in which the future of aluminum is uncertain at the present time is rail and marine equipment. A considerable amount of aluminum metal was used in passenger and cruise ships, but obviously this market is limited. Similarly, there has been a drastic slowdown in the construction of cargo ships. Tankers carrying gas use aluminum in the construction of cyclindrical gas containers, and there is a widespread use of aluminum in small pleasure craft. It seems that an increasing amount of aluminum is being used in underground and short-range surface rail systems, particularly in side paneling and interior accessories. The role of aluminum is also assured in the continuing development of container technology.

Aluminum has found one of its most dynamic applications in packaging, for example, as foil containers used for all-purpose frozen foods, aluminum cans (which are displacing tin cans), and aluminum caps for various types of consumer-goods containers. Aluminum is also rapidly finding favor in the manufacture of aerosol cans. Finally, in domestic and office appliances, aluminum consumption up to now has been modest. About 90% of the total metal content of consumer durables in the early 1970 consisted of steel, while aluminum had only 5%. Given the efficacy of plastics in these applications, it is considered unlikely that aluminum will make any substantial gains here in the near future, although given the relatively high energy intensity of plastics, sharp rises in the price of oil might change this picture.

Table 5-2
Some Physical and Mechanical Properties of Aluminum and Copper

Property	Units	Aluminum	Copper
Melting point	Degrees Celsius	660°C	1083°C
Density	Grams/square centimeter	2.70 g/cm^2	8.96 g/cm^2
Electrical resistance	Ohms/centimeter at 20°C	2.65 Ω/cm°C	1.67 Ω/cm°C
Thermal conductivity	Calories/square centimeter/centimeter/degrees Celsius/second	0.53 Cal/cm^2/cm/°C/s	0.94 Cal/cm^2/cm/°C/s
Modulus: minimum	Pounds/square inch	9×10^6 lb/in^2	10×10^6 lb/in^2
average	Pounds/square inch	10×10^6 lb/in^2	16×10^6 lb/in^2
maximum	Pounds/square inch	11×10^6 lb/in^2	28×10^6 lb/in^2
Average modulus/density	Inches	102×10^6 in	49×10^6 in
1973 price in United States	Cents/pound	25 ¢/lb	59 ¢/lb
Tensile strength (drawn)	Pounds/square inch	34,000 lb/in^2	60,000 lb/in^2
Tensile strength/density	Inches	351,000 in	191,000 in
Maximum tensile strength attained	Pounds/square inch	88,000 lb/in^2	185,000 lb/in^2
Maximum yield strength attained	Pounds/square inch	78,000 lb/in^2	160,000 lb/in^2
Maximum theoretical tensile strength	Pounds/square inch	500,000 lb/in^2	750,000 lb/in^2

Elasticities

Our next step is to review the concept of elasticity in the context of the theory of demand. As most readers already know, a demand curve shows the demand for a commodity given its price. What is important now is that *market demand curves* are an aggregate of individual demand curves and that basically these individual curves need not resemble each other. Figure 5-1 shows portions of two individual demand curves aggregated into a market demand schedule.

The aggregation here is straightforward and takes place horizontally. The value of Q in figure 5-1(c) is simply the sum of q_1 and q_2 in figures 5-1(a) and 5-1(b), respectively, given a value of the price. In particular, note the elasticities of the individual demand curves. In figure 5-1(a) a decrease in price from p_1 to p_2 causes a relatively small increase in demand from q_{11} to q_{12}; while in figure 5-1(b), the increase in demand for the same price decrease is much larger: $q_{22} - q_{21}$. For illustrative purposes, let us take $p_1 = 100$, $p_2 = 99$, $q_{11} = 1900$, $q_{12} = 1910$, $q_{21} = 1900$, and $q_{22} = 2000$. In this case we have a 1% *decrease* in price (= 100 - 99/100 × 100%) leading to a 0.526% (1910 - 1900/1900 × 100%) increase in quantity in figure 5-1(a).

We can now calculate one version of elasticity: the arc elasticity. In all cases elasticity is defined as the percentage change in quantity divided by the percentage change in price or, if the reader prefers, the percentage change in quantity resulting from a 1% change in price, and since in the normal case these changes take place in the opposite directions, the *price elasticity of demand* is defined as being negative. Thus in figure 5-1(a) the elasticity is 0.526/-1 = -0.0526. This is greater than minus one (-1), indicating that a 1% decrease in price is not compensated for by an increase in demand, and thus total revenue (*pq*) falls. Specifically it goes from $p_1 q_{11}$ (= 190,000) to $p_2 q_{12}$ (= 189,090). Over the stretch, or arc, on which the elasticity has been calculated, demand is *inelastic.* In contrast, in figure 5-1(b), elasticity is 5.26/(-1) = -5.26, and the demand is *elastic.* The reader can check what this means for the change in revenue.

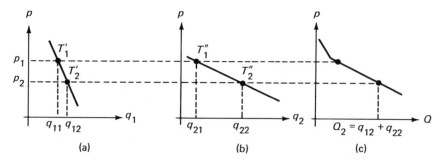

Figure 5-1. Typical Demand Curves and Their Aggregation

We now face a slight problem due to calculating the elasticity over an arc rather than at a point. If instead of a price decrease we had a price increase, say, from 99 to 100, then we get a different value of the elasticity. The issue here can be clarified by writing the formula for the arc elasticity.

$$e = \frac{\Delta q/q}{\Delta p/p} = \frac{(q_{11} - q_{12})/q}{(p_1 - p_2)/p} = \frac{q_{11} - q_{12}}{p_1 - p_2} \cdot \frac{p}{q}$$

Notice that regardless of whether we have an increase *or* decrease in the price equal to $p_1 - p_2$ (or $p_2 - p_1$), there is no change in the *absolute value* of $q_{11} - q_{12}$. It is p and q, or p/q, that changes, depending on whether we have an increase or a decrease in the price. For example, if the price decreases, we have $p/q = p_1/q_{11} = 100/1900 = 0.0526$ in figure 5-1(a). In the same figure, with an increase in price, we would have $p/q = p_2/q_{12} = 99/1910 = 0.0518$. This result leads to different value of the elasticity over the same arc. At this point the reader should go through the same type of calculation using the data in figure 5-1(b).

The way this discrepancy is handled is by making p and q an average of the two prices and quantities. Thus arc elasticity in figure 5-1(a) becomes:

$$e = \frac{q_{11} - q_{12}}{p_1 - p_2} \cdot \frac{(p_1 + p_2)/2}{(q_1 + q_2)/2} = \frac{q_{11} - q_{12}}{p_1 - p_2} \cdot \frac{p_1 + p_2}{q_1 + q_2}$$

Using the numercial values given above, the arc elasticity over $T_1' - T_2'$ in figure 5-1(a) is:

$$e = \frac{-10}{1} \cdot \frac{(199)}{3810} = -0.5223$$

The reader should now calculate the arc elasticity over $T_1'' - T_2''$ in figure 5-1(b). The following should also be observed: Elasticities are often expressed in absolute values. Thus an elasticity of -0.5223 is simply called 0.5223, the understanding being that in the case of a normal demand curve, the reader understands that demand changes are in the opposite direction from price changes. What is important to remember here is that an elasticity that is smaller than unity, or greater than minus unity if we are observing signs (for instance, 0.75 or -0.75), signifies that a given percentage change in price causes a smaller percentage change in quantity. The significance of all this for revenue is that when the demand is inelastic, a decrease in price means that although demand might increase, it will not increase enough to prevent a fall in revenue. Analogously, an increase in price when the demand is inelastic means a percentagewise decrease

Table 5-3
Dynamic and Simple Price Elasticities for the Most Important Minerals

	Dynamic Price Elasticities		Simple Price Elasticities[a]		
	Short Run[a]	Long Run[b]	Low	High	Substitutability and Substitutes
Aluminum-Bauxite[c]	−0.13	−0.80			Yes (Copper, steel, wood plastics, titanium)
Chromite	−0.10	Elastic			Yes (Nickel, vanadium)
Cobalt	−0.69	−1.71			Yes (Nickel)
Copper	−0.20	−2.00			Yes (Aluminum, plastics)
Iron ore			−0.10		
Lead			−0.10	−0.30	Yes (Rubber, copper, plastics, zinc)
Manganese ore				−0.10	
Molybdenum	Inelastic				Yes (Vanadium, tungsten)
Phosphate rock				−0.10	
Platinum-palladium	Almost zero	−0.9			Yes (Gold)
Tin		−0.8			Yes
Tungsten	−0.15	−0.30			Yes (Molybdenum)
Zinc	−0.55	−0.70			Yes (Aluminum, plastics)

[a]Minimum

[b]Maximum

[c]Pindyck (1977) estimates the short and long run elasticities to be −0.20 and −1.00 respectively.

Note: In estimating these elasticities, a dynamic demand curve was first specified. If this gave no results, a nondynamic formulation was tried. For a complete discussion of these matters, as well as a listing of some elasticities of supply, the reader should refer to Banks (1977).

in demand that is less than the percentagewise price rise. Consider, for example, the situation with oil. Even when the price increased by almost 400%, demand fell only marginally, and producer revenues were not only maintained but increased drastically.

We can now give some price elasticities of demand for various mineral commodities, and also say something about their substitutability. The reader should remember that these elasticities are only approximate, although they do provide a good indication of whether demand for the mineral is elastic or inelastic.

"Dynamic" Demand Curves

The previous section discussed cases of inelastic and elastic demand. It could happen, however, that we might have both, depending on our time scale. Consider the diagram in figure 5-2.

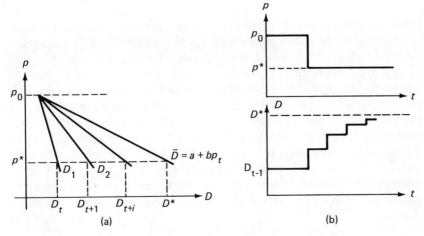

Figure 5-2. Dynamic Demand Curves Showing an Increase in Demand Following a Decrease in Price

If, for example, there were a decrease in the price of aluminum metal from p_0 to p^*, as shown in figure 5-2(a), what would it mean to the industrial purchasers of aluminum? To begin, their plants are designed for a certain optimal output—that is, to process a certain amount of aluminum per time period. By running their machines faster and more hours per day and obtaining more overtime from their employees, they could increase their aluminum requirements somewhat, but obviously their flexibility is constrained in the short run by the design of existing facilities. Moreover, in the case of manufacturers of such things as transport equipment, aluminum plays such a small role in the final product that there is no great incentive to go to the trouble and expense of altering the design of various components even if the price decrease were fairly substantial. A large decrease in price, however, might cause many users of this metal to increase their inventories, particularly if they thought that at some time in the future, prices might increase again.

Next consider the medium to long run. If it appeared that the decrease in the price of aluminum were permanent or even that the price of competitive materials such as copper, plastics, or steel might increase, aluminum purchasers would be compelled to take this situation into cognizance when making their investment plans. The design of engine blocks could be altered to the extent that additional aluminum parts were incorporated, and house builders would specify more aluminum wiring. These measures, however, would take time; and this time would usually be measured in years rather than months. In other words, the key factor in the adjustment of aluminum consumption to changes in price is the size of the stock of aluminum-using machinery and the rate of its depreciation. It is dubious, for instance, that a sharp fall in the price of a metal would invoke the immediate establishment of many new industries using that metal.

Some of the preceding observations can be translated directly into algebra, using figure 5-2 as a reference. If we begin with price p_0 and demand D_{t-1}, a fall in price to p^* will result in a long run, or equilibrium, demand of D^*. But this position will be approached only gradually. In the first period after the change in price, the intention of buyers is to increase demand by the amount $\lambda(D^* - D_{t-1})$. For instance, if λ were $1/4$, $D^* = 200$, and $D_{t-1} = 100$, the increase in demand would be 25 during the initial period, bringing the total demand to 125. During the next period the increase in demand would be $\lambda(D^* - D_t)$, which would come to $\frac{1}{4}(200 - 125) = 18.75$. Total demand is thus $100 + 25 + 18.75 = 143.75$. To compute total demand after the next period, the same procedure can be repeated, but at this point we will go over to a formalization of our exposition. Thus we have:

$$\Delta D_t = D_t - D_{t-1} = \lambda(D^* - D_{t-1})$$
$$\Delta D_{t+1} = D_{t+1} - D_t = \lambda(D^* - D_t)$$
$$\Delta D_{t+2} = D_{t+2} - D_{t+1} = \lambda(D^* - D_{t+1})$$

and so on. Confining ourselves to 2 periods, we can write:

$$
\begin{aligned}
D_{t+2} &= D_{t+1} + \lambda(D^* - D_{t+1}) \\
&= D_t + \lambda(D^* - D_t) + \lambda[D^* - D_t - \lambda(D^* - D_t)] \\
&= D_{t-1} + \lambda(D^* - D_{t-1}) + \lambda[D^* - D_{t-1} - \lambda(D^* - D_{t-1})] \\
&\quad + \lambda\{D^* - D_{t-1} - \lambda(D^* - D_{t-1}) - \lambda[D^* - D_{t-1} - \lambda(D^* - D_{t-1})]\}
\end{aligned}
$$

The manipulation involves continuous substituting from the previous set of relations in order to get everything in terms of D^* and D_{t-1}. Now the last expression can be simplified to:

$$D_{t+2} = D_{t-1} + \lambda(D^* - D_{t-1}) + \lambda(1 - \lambda)(D^* - D_{t-1}) + \lambda(1 - \lambda^2)(D^* - D_{t-1})$$

If we were to use the above numerical example, we would have:

$$
\begin{aligned}
D_{t+2} &= 100 + \tfrac{1}{4}(200 - 100) + \tfrac{1}{4} \cdot \tfrac{3}{4}(100) + \tfrac{1}{4} \cdot \tfrac{3}{4} \cdot \tfrac{3}{4}(100) \\
&= 100 + 25 + 18.75 + 14.0625 = 157.8
\end{aligned}
$$

In other words, with the new equilibrium demand at 200, at the end of *three* periods (since we began at $t - 1$), demand has increased to 157.8. The proof that it will eventually reach 200 is simple, although reaching this level takes a great deal of time—infinity to be exact, as might be inferred from the development shown in figure 5-2(b). What this proof involves is simply the extending and rearranging of a series of the type shown above. Thus we get:

$$D_{t+n} = D_{t-1} + \lambda(D^* - D_{t-1}) + \lambda(1 - \lambda)(D^* - D_{t-1}) + \cdots + \lambda(1 - \lambda)^i (D^* - D_{t-1})$$
$$+ \cdots = D_{t-1} + \lambda(D^* - D_{t-1}) [1 + (1 - \lambda) + (1 - \lambda)^2 + \cdots]$$

The series in the parentheses can be summed in the usual way. Taking $n \to \infty$, we get:

$$D_{t+n} = D_{t-1} + \lambda(D^* - D_{t-1}) \cdot \frac{1}{1 - (1 - \lambda)} = D_{t-1} + D^* - D_{t-1} = D^* \quad \text{Q.E.D.}$$

Now for elasticities. In figure 5-2 the original price is p_0, and the long-run demand curve is $\bar{D}$. As shown in the figure, if demand adjusted immediately to a fall in price to p^*, we would move to D^*. Instead, in the very short run, with price p^*, demand becomes only slightly larger than its initial value D_{t-1}. Specifically, it reaches D_t. Later it arrives at D_{t+1}, still later at D_{t+2}, and as time goes by we get an asymtotic movement to D^*. The curves D_1 and D_2 are typical intermediate demand curves, and it is possible to draw any number of demand curves to the right or left of these curves.

The law of adjustment for this model is $D_t = D_{t-1} + \lambda(\bar{D} - D_{t-1})$, where $\bar{D}$ is an equilibrium level of demand, and $0 < \lambda < 1$. When $D = \bar{D}$, or the actual value of demand at time t is equal to what is perceived as the equilibrium or desired value at time t, the system is stationary. In figure 5-2 long-run demand is related to price by $\bar{D} = a + bp_t$, and this is put into the adjustment equation. We then get $D_t = \lambda a + (1 - \lambda)D_{t-1} + \lambda bp_t$. When $D_t = D_{t-1} = \bar{D}$, we have an equilibrium, which can be verified by putting this condition into the previous equation. This takes us back to the long-run demand curve $\bar{D} = a + bp_t$, which functions conceptually as an equilibrium demand curve. (*Note:* When $p_t = p^*, \bar{D} = a + bp^* = D^*$.)

With this settled, we can distinguish between long- and short-term elasticities. With an equation such as $D_t = \lambda a + (1 - \lambda)D_{t-1} + \lambda bp_t$, we have a long-run equilibrium when $D_t = D_{t-1}$, and thus we get:

$$D_t = a + bp_t$$

Accordingly the long-run elasticity taken at the point (p', d') is:

$$e_L = \frac{\partial D_t}{\partial p_t} \cdot \frac{p'}{d'} = b\frac{p'}{d'}$$

On the other hand, the short-run elasticity is taken directly from $D_t = \lambda a + (1 - \lambda)D_{t-1} + \lambda b p_t$, and with $0 < \lambda < 1$, is smaller than e_L, as expected.

$$e_S = \frac{\partial D_t}{\partial p_t} \cdot \frac{p'}{d'} = \lambda b\frac{p'}{d'}$$

This is all very well, but as will be shown later, the emphasis in the analysis must be guided away from *flows* such as D and toward *stocks*. In addition, some notice must be taken of expectations. Not only are we dealing with a concept of demand whose mechanics is considerably different from that implied in figure 5-1 but also the structure of expectations on the part of market participants might be such as to reverse some conventional results on elasticities. For instance, suppose demand for an item takes into consideration not only the present but the expected price. For instance, we might have:

$$D_t = f(p_t, p_{t+1}^e) = a_0 + a_1 p_t + a_2 p_{t+1}^e \qquad a_1 < 0, a_2 > 0$$

Notice the specification of a_2, which indicates than an expected increase in the price of a product in the coming period will lead to an increase in present demand. Next we need to specify a scheme for expectations. To keep things simple, let us take these as extrapolative, which means that $p_{t+1}^e = p_t + \phi(p_t - p_{t-1})$. Thus if $\phi > 0$ and we have an increase in price from period $t - 1$ to period t, a higher price will be expected in the following period. Using p_{t+1}^e in the previous equation gives us:

$$D_t = a_0 + [a_1 + a_2(1 + \phi)]p_t - a_2 p_{t-1}$$

Given the signs of a_1 and a_2, it is apparent that $a_1 + a_2(1 + \phi) \gtreqqless 0$, depending on the magnitudes of a_1, a_2, and ϕ. Under the circumstances, we could end up with a positive elasticity of demand. Naturally, we would not expect this arrangement to prevail all the time. Instead, ϕ would probably be a variable and itself a function of such things as past prices, inventory levels, and so on.

The reader will find some typical elasticities in table 5-3.

Supply

We can now continue our discussion of the supply of aluminum. As we know from elementary economic theory, producers change supply in response to changes in the price of the product they are producing, as well as changes in the cost of the factors they use to produce these products. Basic economic doctrine says that if, for example, we have an increase in the price of a product due to an increase in demand, producers will increase its supply by first increasing their input of the variable factors of production such as labor, raw materials, and energy; and if they think the increase in demand is a permanent phenomenon, they will expand their production facilities. This could take place through, for example, increasing the scale of existing facilities or scrapping or selling some existing equipment and replacing it with more modern equipment. Thus we can think in terms of a short-run supply curve that applies to a given capital structure, with changes in output being brought about through changing the input of variable factors; and also a long-run supply curve on which the capital structure is optimized in the sense that any given output is being produced at the lowest possible cost due to capital and labor (and other production factors) being combined in "optimal" proportions.

Next we can turn to the matter of inventories, beginning our analysis with an upswing in the business cycle that causes an increase in the demand for various metals. As outlined above, the production of these metals will increase, but slowly at first since producers must ascertain future prospects before they order expensive pieces of capital equipment. In addition, the labor market might be such that it is difficult to obtain additional personnel, or the existing capital equipment might not provide a great deal of scope for increasing production simply by increasing the input of one or more variable factors. In this type of situation a large part of the responsibility for satisfying the increase in demand must fall on inventories. Thus the increase in demand would initially tend to reduce inventories relative to demand (or consumption of the particular metal), and the rise in price that we observe is a direct concomitant of this particular phenomenon. This explains, incidentally, why the fall in production of copper resulting from the partial destruction of production facilities in Zaire, together with the maintenance of a fairly strong world demand for copper, did not lead to sharp price increases for this metal: existing stocks of copper were still regarded as excessive, and thus the gap between *current demand* and *current production* was easily filled by reducing inventories. This topic will be gone into much more thoroughly later in this chapter, and in the next.

We can now examine the type of model required to obtain a satisfactory amount of explanatory power for a situation involving inventories. Everything considered, the simple supply-demand model of the elementary textbooks must be replaced by one that introduces inventories explicitly. That is, instead of the customary scheme shown under column A, we need (1), (2), and (3) under column B, together with a relationship similar to one of the four possibilities given as (4). In this presentation s is supply, d demand (for current consumption), and p price.

A		B	
(1) $s = s(p)$		(1)	$s_t = f(p, \ldots)$
(2) $d = d(p)$		(2)	$d_t = g(p, \ldots)$
(3) $s = d$		(3)	$s_t - d_t = \Delta I$ (I = inventories)
		(4) (a)	$p_t = p_{t-1} + h(\Delta I)$
		(b)	$p_t = p_{t-1} + w(\Delta d / \Delta I)$
		(c)	$p_t = p_{t-1} + \theta(k\Delta d - \Delta I)$
		(d)	$p_t = p_{t-1} + \lambda(I^* - kd)$

Under column B, θ, λ, and k are parameters, while f, g, h, and w indicate functions. What is being done here is to take the first step in reaffirming the important but generally overlooked work of Clower (1954) and Bushaw and Clower (1957). Industrial raw materials are not just flow commodities, in which case the model listed under column A would suffice, but stock-flow commodities, and the analysis must treat this fact explicitly.

This brings us to some definitions: A *stock* is a quantity measured at an instant of time while a *flow* is the time rate of change of a variable. Capital (such as machinery) is stock, while investment (which is the change in the capital stock per period) is a flow. In other words, a flow is characterized by a time dimension as well as a time reference. Similarly, we might distinguish between stock, flow, and stock-flow markets. A stock market is a market concerned with the holding of a commodity, in contrast to the current production and consumption that characterizes a pure flow market. An example of the first of these is a market for the paintings of Rembrandt; and while the second type of market is becoming rarer due to the widespread application of, for example, refrigeration, a candidate for this category of product would be fish products such as crayfish or agricultural products like watermelon, especially if our "period" were made long enough.

On the other hand, the market for a metal such as aluminum or copper would be a typical stock-flow market: not only is there a demand for these metals as current inputs but there is also a stock or inventory demand governed by an entirely different set of considerations. The model shown under column B, while intended to give some idea of the kind of factors that influence price formation on the market for an industrial raw material, cannot be employed in its present form. It does not, for example, yield a direct analogy to the simple but highly effective graphical apparatus that is associated with model A. Moreover, there can be some embarrassing dynamic effects associated with model B that given an arbitrary displacement of the system, can cause endogenous variables to move *away from* rather than *toward* a new equilibrium.

Some aspects of the present world supply situation on the aluminum market can now be examined. As far as I can tell, at the present time the price of bauxite is steady, reflecting the willingness of producers to continue supplying ore at present prices. There is also a satisfactory supply of aluminum metal, although most predictions are for a tightened supply by the first part of the 1980s. Here it should be noted that although, on the basis of an elementary trend analysis, the demand for aluminum ingot should expand at a rate of about 7% a year, it would be exceptional if, given the present rate of investment in aluminum smelters, ingot output could increase by more than 6% a year. Figure 5-3 shows the *price-shipment* situation for the United States from 1970 to the beginning of 1978.

In case the reader is concerned about the sharp dip in shipments shown in this figure, this situation originated on the demand side of the market. But since prices are "administered," as will be explained later, and given the low price elasticity of demand of aluminum, there was no economic reason for producers to lower their prices at that time.

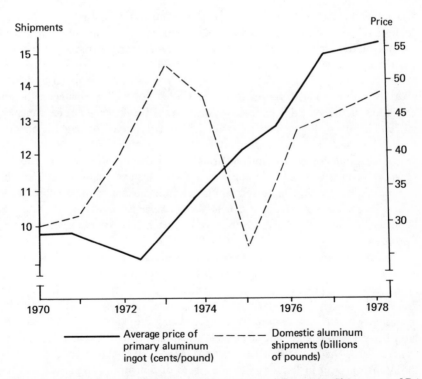

Figure 5-3. The Producer Price for Aluminum, and Domestic Shipments of Primary Unalloyed Aluminum Ingot in the United States (1970–1978)

As already made clear, one of the main future sources of uncertainty for aluminum producers is the cost of electric power. Even though ingot prices are on an upward trend at present, profitabilities could be kept down by accelerating energy costs. A somewhat different problem faces the sector handling semi-fabricated products. Here the source of supply reticence is the memory of recent excess capacity, particularly in the early 1970s but also in 1975. Clients of this sector, and especially buyers of heart-treatable sheet and forgings, are experiencing both higher prices and longer delivery times. The position of aluminum fabricators as a group is that they are not going to construct new facilities until the long-run demand pricture is clarified; and first and foremost, in the near future they are going to be concerned with short-run profitability. In practice, this will mean more modernization of existing facilities with the emphasis on the introduction of labor-saving technology and a general lack of interest in expanding capacity. In the United States, this strategy appears to yield returns to investment as high as 25% in a few cases, compared to the 10–12% average return to new capacity.

Japan and Australia should also be referred to in the context of the present discussion. As pointed our earlier, the increase in the cost of electric power means that Japan is on the way to losing its comparative advantage in the production of aluminum metal and aluminum products—an advantage it gained because of the technological leadership of its capital- and energy-intensive production apparatus. The increase in world oil prices drove the price of electric power in Japan up from 3.5 yen/kWh to 8.5 yen/kWh, as compared to 4 yen/kWh for Europe, and 2 yen/kWh for the United States. Together with the appreciation of the yen, this increase means that the breakeven point for Japanese smelters is now in the vicinity of 1950 U.S. dollars/ton as compared to a world average of 1490 U.S. dollars/ton. (Of course, appreciation of the yen could be compensated for by subsidizing firms out of the "profits" accruing to the Japanese economy because of appreciation. For instance, appreciation of the yen means that users of imported inputs priced in dollars or other depreciating currencies realize a gain; which is true also in Japanese and German and Swiss importers of energy relative to energy importers in, for example, Sweden, France, Australia.)

Japan can therefore be expected to increase its imports of both aluminum ingot and aluminum fabricates and also make a determined effort to locate more of its aluminum-processing capacity overseas. The all-powerful Ministry for Trade and Industry has directed the Japanese aluminum sector to consolidate its operations (that is, reduce output); and processing sites in countries blessed with both bauxite and energy, and if possible cheap labor, are being investigated at a frenetic pace. The largest Japanese overseas investment in aluminum at present is the Asahan project in Indonesia on the island of Sumatra. This project calls for the construction of a 225,000 ton/year aluminum smelting plant together with a 513-megawatt hydroelectric power plant. Japanese equity will amount to

37.5%, and the $2.16 billion project is scheduled to be completed by 1984. The Mitsui Trading Company has also taken a substantial equity position with AMAX Mining of the United States, whose aluminum producer and fabricator, Alumax, has the largest growth plans in the United States: Announced intentions are to expand capacity by 115% up to 1982, which means an increase of 253,000 tons as compared to 217,000 tons for all other U.S. producers.

Australia also has a place in this drama. At present there are three smelters in that country, and the very favorable economics of electricity generation on the eastern coast of Australia as well as the huge bauxite supplies of Australia have caused a number of international groups to start thinking in terms of sponsoring increased smelter capacity. A proposed operation in New South Wales would involve an equity distribution of 50% of AMAX, 45% to Mitsui, and 5% to Nippon Steel. Capacity would be 200,000 tonnes/year, and initial capital outlays would amount to $500 million, which, among other things, would create about 1000 jobs for Australians on the construction side of the venture. What it comes down to is that with this type of operation, Australia has begun, and will probably continue to an even greater degree, to export items to Japan having as high a technological content as the products Australia has been importing from Japan for at least the last decade.

Just what this will mean in the long run for the economies of Japan and Australia, as well as the world economy, is difficult to say; but the opinion here is that for various reasons, most of which are well known and need not be repeated here, Japan is ideally suited for the manufacture and export of goods with a high technological content, at least at the present time, and if that country is prevented from carrying out this function because of its unfavorable energy situation, then the entire world economy will be adversely affected.

Prices and Pricing

The final topic in this chapter has to do with producer and free market pricing, an important but neglected topic as far as the nonfuel minerals markets are concerned. In continuing the exposition begun earlier in the chapter, the first point to stress is the cloud of uncertainty that hangs over all market participants. Regardless of their efficiency, producers only *think* they can produce and deliver certain amounts to their customers by or on a certain date. Something can always happen to frustrate their plans: strikes, breakdowns, wars, and so on. A similar situation faces consumers. They believe that they will require a certain amount of ore, concentrate, or refined products at a certain time, but it could happen that they require more or less. As a result a model that does not make allowances for discrepancies in plans or intentions in the broadest possible sense will not take us far in understanding the real world, and once again it turns out that a key component of such a model is inventories.

Almost all categories of seller and buyer hold inventories, either planned (or desired) inventories or unplanned (that is, undesired). But selling and buying practices may differ at various stages of processing. For example, ores and concentrates are usually traded via face-to-face negotiations between producers and consumers or their agents, and the problems that arise here have been reviewed earlier in this book (see particularly the last section of the previous chapter). Of course, a situation could exist in which these particular market participants are identical since the high degree of market integration within many minerals industries means that a great deal of selling and buying reduces to book-keeping entries between divisions of the same firm.

The sales of these products often take place using the medium of *forward contracts.* This type of contract covers the sale of an item that will be delivered in the future at some mutually agreed on fixed price. Specifically, the forward contract, which is conceptually an instrument of the forward market, involves physical delivery under a given set of conditions, and at a given time or within a given period, and as such it should be distinguished from the *futures market,* where delivery takes place only in the minority of cases. (This matter will be cleared up in the next chapter.) At the same time it must be emphasized that the forward market is not a market in the geographical or institutional sense, but rather an arrangement.

The same holds true for the *spot* or "cash" market, which involves delivery in the immediate future. The expression "immediate," however, is flexible in that in the market for iron ore, for example, it can mean up to one year. Also where this commodity is concerned, forward contracts designated long-term contracts have run up to 20 years, and among their clauses we often find provisions calling for periodic renegotiation of both quantities and prices.

As stated, the preceding applies to such things as iron ore, alumina, copper concentrates, and "blister," and other ores and semiprocessed materials. It also applies to refined products; but a complete examination of these items requires some important new concepts. Moreover, in the economic—as distinguished from the engineering or commercial—literature, most of the discussion refers to refined products, although this is not always made clear. The reason for focusing on these products is due to their high degree of homogeneity, which may not be true in the case of, for example, ores, concentrates, and even fabricated products.

This homogeneity means that they can be sold on commodity exchanges, where various characteristics must be guaranteed, or even on the basis of listings in trade journals. As a good example of this type of commodity, we find that the standard unit of nickel is a cathode sold by International Nickel and Falconbridge in Canada. Other sellers of primary nickel compare the quality of their products with these cathodes. An important set of prices quoted in the trade journals are known as *producer* or *posted* prices. These are "industry" or "official" prices and are the prices at which the largest amount of the metal is sold. But in addition to these administered prices, there is a small free market for primary nickel and nickel scrap. The primary nickel sold on the free market in Western Europe sometimes originates in the U.S.S.R. and other centrally planned economies and is purchased by *merchants* or *dealers* who resell at prices that depend on supply and demand. In other words, if buyers underestimated their requirements and did not purchase enough metal in the forward market, they reduce inventories or turn to the free market or both.

Still looking at nickel, during 1969-1970 the supply of this product by the large producers in the market economies could not meet demand, and consequently buyers were very active in the free market—"free" in the sense that it was very sensitive to current supply and demand pressures. At that time free market prices commanded a large premium over the producer price; and in 1969, during the long strike in the Canadian mines, the British Steel Corporation doubled its purchases from the free market, at times paying prices that were six or seven times the producer price. During the last 7 years producer and market prices have been fairly close, with the market price averaging 16% above the producer price in 1974 and 9% below in 1975.

The Price of Aluminum

As indicated earlier, the situation for aluminum is complicated somewhat by the high degree of integration in the aluminum industry. Thus it has been so in recent years that 80% of primary aluminum production goes from the primary producer to a subsidiary producing semifabricates at a book-keeping price that may or may not be related to the market price.

The basic world producer price is set or administered by the largest exporter, Alcan. Other exporters mostly follow Alcan's lead, although there are undoubtedly some discussions between the major producers. Producer prices quoted within countries can differ from the world price; and for reasons that will be given later, the introduction of an aluminum contract on the London Metal Exchange will mean that to a certain extent, producers will lose some control over the price of aluminum. At the present time, as with nickel, there is a dealer's or free market for aluminum, and the price or prices on this market are regularly quoted in trade journals. It could eventually happen, however, that outside the United States the principal free market price for this metal will be that quoted on the London Metal Exchange. Another point that should be noted here is that traditionally, in the aluminum industry it has been the larger companies that rigidly held to producer prices, while when demand "softened," the smaller producers resort to discounts almost immediately.

Table 5-4 shows the producer price in a number of countries together with the world (export) price for the period 1970-1974. All prices have been converted to U.S. cents per pound using average exchange rates. Prior to this period, during the 1960s up to 1968, prices changed very little, and in general there was

Table 5-4
Some Producer Prices of Aluminum in Five Countries and the World Price of
Aluminum (1970-1974)

Country	1970	1971	1972	1973	1974
France	27.1	27.7	30.1	33.2	35.4
Germany (F.R.)	28.3	29.8	30.7	36.7	41.9
United Kingdom	27.9	28.4	26.7	27.1	34.3
Japan	28.0	26.5	28.7	34.5	46.8
United States	28.7	29.0	26.4	25.0	34.1
World Price	27.9	28.0	26.0	27.3	36.3

a balance in supply and demand over most of the world. This balance was reflected to a considerable extent by the proximity of producer and free market prices and their tendency to follow the same trend. Many observers are convinced that this market stability was attributable to the high degree to integration prevailing throughout the world economy.

After 1970 the picture changed. In 1971 and 1972 most producers had excess capacity, and there was a pattern of declining earnings and profits right across the industry. In late 1972, 1973, and even up to the first part of 1974, with the war in Vietnam over and the world economy gaining strength as a result of various currency realignments, the demand for aluminum increased, as did industry prices and profits. This is a fruitful period for economists to study since it offers a chance to see how a real—as opposed to a textbook—market functions. In the first part of this period, a number of new sellers came on the market, selling considerable quantities at free market prices. Some aluminum also reached the West from Communist countries, and from one end of the industry to the other, substantial discounts and markdowns became the practice. Even the fabricating branches of major producers had to put pressure on the suppliers of metal within the same firm to reduce prices; and it was estimated that in the first weeks of 1972, the free market price was 20% lower than the producer price.

The world economic upswing that began in 1972 meant that a balance was established on the market for aluminum metal by the beginning of 1973. At that time consumption was 18% above the 1972 average, and free market prices rose considerably above producer prices, even though producer prices also increased during the same period. The explanation for this price increase could not be confined to rising demand but was also a function of the slowing down in capacity expansion that was initiated by most aluminum producers in 1971-1972 as well as strikes and electricity shortages. It is also probable that prices would have gone much higher had it not been for the U.S. Strategic Stockpile's releasing metal to the market and, in addition, a steady flow of aluminum from Eastern Europe. Some of these price trends are visible in figure 5-4, which displays pro-

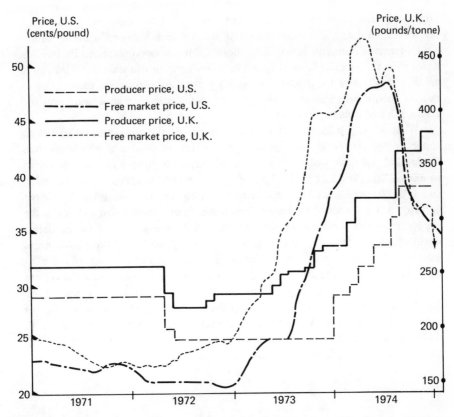

Figure 5-4. Producer and Free Market Prices for Aluminum Metal in the United States and the United Kingdom (1971–1974)

ducer and free market prices in the United States and the United Kingdom between 1970 and 1975.

With the advent of the "energy crisis," free market prices plunged, and like most industries, the aluminum industry entered the doldrums. Some experts believe that the industry is on its way out of this sad state, but the opinion here is that if this is so it is because they are not expanding capacity at a rate that aluminum consumers would judge desirable if they gave the problem some thought although, if the price of oil increases by steps of 25% every few years, as was the case in 1979, the world economy will stay weakened to the point that the demand for aluminum could easily repeat one of its previous collapses.

One final point should be made here. The reader examining figure 5-4 might wonder just why the producer and free market price of a metal should differ to the extent that they often do. In particular, we might ask just why, when the market price is higher than the producer price, producers do not adjust up the producer's price right away. After all, isn't this what the elementary textbooks say they should do? The short answer to this riddle is that producers are thinking in terms of long-run rather than short-run profit maximization. By holding down the average price level, they inhibit investment in this industry. Thus, later on they might avoid a situation in which there will be so much capacity as to exert a downward pressure on the price of metal as well as an upward pressure on the price of industry inputs.

What all this might result in is a situation of the type that existed with copper in the United States a decade ago. Keeping the producer price lower than the market price for a number of years inhibited investment in this industry to an extent that President Richard Nixon felt compelled to establish a commission under the leadership of Professor Henrik Houthhaker to investigate the copper industry. Unfortunately, Professor Houthhalker's erudition failed to impress the copper magnates, and eventually he returned to the glories of the academic world. To show that his time was not completely wasted his commission did issue a preliminary report, which Houthhaker summarized in a leisurely chat at Duke University in March 1970. Among other things he diagnosed the copper industry as suffering from a chronic malfunction, although, as expected, he could not propose a cure. As things have worked out though, a number of copper producers in the United States have recently decided to break ranks with tradition and sell their product at the free market price, but whether this will solve anything in the long run remains to be seen.

Appendix 5A
The Substitutability of
Aluminum for Copper. and
an Econometric Comment
on Elasticities

5A-1.* The question of the substitutability of aluminum for copper has come up at several points in the preceding discussion, but there are still a few observations that deserve to be made.

Figure 5A-1 is a diagram of the ratio of the price of copper to the price of aluminum over the period 1945-1973. With the exception of the Korean War, this period saw the industrial world go through the post-World War II reconstruction followed by a kind of golden age that *may* have ended with the Vietnam War but that definitely ended, at least for the time being, with the October War of 1973 and the oil price increases. The figures used to construct this diagram are global averages.

On the basis of the obvious trend in this price ratio, some people might surmise that aluminum is increasing in importance relative to copper. (although, in truth, somewhat more evidence might be required). Moreover, it seems clear that during the period 1956-1968 there were times when copper was a spectacularly good buy in comparison to aluminum.

But it was precisely during this period that the consumption of aluminum actually accelerated—both absolutely and in relation to copper. Note also that in line with the discussion in this chapter, the "fall" in the price of copper relative to aluminum lasted over a very long time. Long enough, in fact, to have a decisive influence on the potential users of these two metals; and on the basis of traditional economic theory, we certainly would have expected, toward the end of this period, to have seen a great deal of substituting of copper for aluminum. But as far as I can tell, this is the reverse of what actually happened.

5A-2. We can now go over to a short review of an econometric examination of these matters. In the body of the chapter we derived the equation $D_t = \lambda a + (1 - \lambda)D_{t-1} + \lambda b p_t$, which if D is actually a function of p, could be put through the standard econometric routine, and the value of the parameters λ, a, and b derived without further ado. [As shown in Banks (1974b, 1977), though, we would probably start out by assuming that D was a function of both price and income instead of just price.) However, in my own econometric work on aluminum, which has just started, this approach has not led to any usable results. The reason is obvious once we look at a plot of aluminum consumption versus manufacturing output as shown in figure 5A-2. As could have been expected a priori,

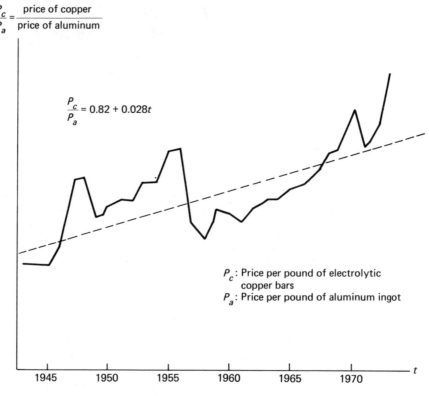

$\dfrac{P_c}{P_a} = \dfrac{\text{price of copper}}{\text{price of aluminum}}$

$\dfrac{P_c}{P_a} = 0.82 + 0.028t$

P_c: Price per pound of electrolytic copper bars
P_a: Price per pound of aluminum ingot

1945   1950   1955   1960   1965   1970 t

Figure 5A-1. The Price Ratio of Copper to Aluminum (1945–1973)

aluminum has a very low price elasticity in the short run—as far as I can tell at the present time, much lower than copper—and I have estimated the following relationship for explaining the aluminum consumption of the OECD countries for the period 1960–1972.

$$\log D = -3.940 + 2.321 \log (\text{GNP}) \qquad R^2 = 0.996$$
$$(48.54)$$

For various statistical reasons, GNP is used here as a proxy for industrial output. The numerical value in parentheses is the t ratio for the coefficient of log (GNP), which happens to be the income elasticity of demand for aluminum.

In looking at this matter from another direction—that of copper consumption—Professor McNicol and various associates have estimated an equation for the consumption of copper in the United States of the following type:

$$C_t = \alpha_1 C_{t-1} + \alpha_2 P_{tc} + \alpha_3 P_{ta} + \alpha_4 X_t + \alpha_5 Z_t + \alpha_6 Z_{t-1}$$

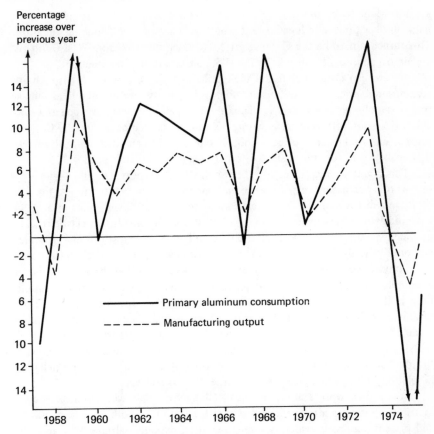

Figure 5A-2. The Cyclical Fluctuation in Aluminum Consumption in Relation to the Percentage Change in Manufacturing Output for the Noncentrally Planned Industrial Countries

Here C is the consumption of copper, P_{tc} the price of copper, P_{ta} the West German price of aluminum deflated by the U.S. wholesale price of durable manufacturers, X_t the index of production of durable manufactures in the United States, and Z the real changes in manufacturers' inventories of durable manufactures.

What we are interested in here is the "substitution term," or P_{ta}, which for some astonishing reason relates the consumption of copper in the United States to the West German price of aluminum. And why West German price? According to McNicol and other model makers, when the U.S. aluminum price was put into the preceding equation, not only did its coefficient fail to be positive but the rest of the equation was radically altered. Thus instead of the U.S. price which, considering that we are examining demand in the United States, would a priori seem to be the correct price, it was necessary to go through the data bank until a price was located that would give a respectable-looking equation. This turned out to be the German price, although imports from West Germany are almost an insignificant part of the U.S. consumption of aluminum.

It was also claimed that the U.S. price and the London (LME) price are administered prices, and because of their inflexibility, are unsuitable as indicators of supply-demand pressures. Thus the German price was advanced as a suitable price. But this solution is wrong because it also happens that the German price is an administered price, although one that has an extra kink or wriggle that enables it to turn unsatisfactory equations into marketable items.

This is not science, of course, but sophistry. Moreover, I would advise anyone who is purchasing any of these equations to pay careful attention to the size of these substitution effects in relation to the size of "own" price effects. In the models that are already on the market, short-run substitution elasticities are as much as three times own elasticities; and long-run substitution elasticities are only trivially larger than short-term elasticities. This by itself means that these models bear no relation to reality since it happens that the substitution of aluminum for copper is a secular phenomenon based on the belief that aluminum is the metal of the distant future due to the well-documented abundance of aluminum-bearing ores, as well as to the technological advances that are being made in producing and using aluminum. These reasons translate into price effects, quite naturally, but not those that can be picked up in the short run with the dull blade of linear regression: Short-run substitution effects are neligible in relation to long run, and the same is true with own effects.

In my own work I am constantly talking to people who purchase econometric output year after year and whose frustration at both its lack of utility and pretentiousness is evidently increasing at an exponential rate. Why then do they not go over to a more scientific method of forecasting commodity prices such as reading tea leaves, or examining the entrails of garden insects?

5-A3. By way of presenting a more solid example of the relations introduced in this chapter, I have fitted the following simple supply equation for world tin output, employing annual data for the period 1960-1974.

$$S_t = 0.8602 \, S_{t-1} + 0.0050 \, p_{t-1} \qquad\qquad R^2 = 0.930$$
$$\quad\;\; (14.01) \qquad\quad (2.88) \qquad\qquad\qquad \text{D.W.} = 2.18$$

The total supply of primary tin is S_t plus the supply (positive or negative) of the U.S. Strategic Stockpile. What the reader should notice here is the very small coefficient (0.0050) of the price term. As for the price elasticity (with respect to price in period $t-1$), this is evaluated at the mean of the values of S_t and p_{t-1} and is:

$$e_{ps} = \frac{\partial S_t}{\partial p_{t-1}} \cdot \frac{p_{t-1}}{S_t} = 0.1299 \qquad \text{Short-run price elasticity}$$

$$e_{pL} = 0.90 \qquad \text{Long-run price elasticity}$$

For some typical price elasticities of supply, the reader is referred to Banks (1977); and for a consumption equation for tin, and a calculation of the demand for tin in the United States, the reader can examine Banks (1972).

6

Prices, Markets, and Inventories

Metal Exchanges and Exchange Pricing

This chapter is a continuation of the previous one, although in parts it will display a slightly higher technical level. However, to begin we need to discuss some aspects of the metal exchanges and exchange pricing.

The prices of some of the most important industrial raw materials are quoted on the two major commodity exchanges, the London Metal Exchange (LME) and the New York Commodity Exchange (COMEX). Of these two, the LME is considered the most important in terms of turnover, physical deliveries, and its influence on the pricing of metal in general. On the other hand, until recently COMEX handled a wider variety of metals and in addition provides facilities for trading hides and rubber. It seems to be true that in the future, these exchanges will be even more important than they are at present since at least some American copper producers intend to price their copper on the basis of exchange prices, while aluminum and nickel contracts are being introduced on the LME.

Despite the inference found in the term "exchange," the buyers and sellers of metals and other products do not usually come to the commodity exchanges to make their deals. They simply agree on a formula for pricing the commodity in which they are interested, relating the price of the commodity to a price or prices on an exchange. One possibility is that the price at which the commodity will be traded will be taken as the average of the "spot" prices prevailing on the LME during the week before the scheduled date of delivery of the commodity. As for these spot (or cash) prices, they are determined by a small amount (by volume) of trading that takes place daily, during a limited time period, on the exchange.

The LME dates from 1882, although various metals had been quoted in London earlier on informal exchanges. During World War II and the immediate postwar period, the government controlled the price of raw materials, and so the exchange was closed; but it reopened in 1953, and since that time a steady increase in activity has been recorded. Once again it should be emphasized that this activity has only a small physical component. In 1968, with the world consumption of copper equal to approximately 7 million tons, approximately 2 million tons of copper "futures" were traded on the LME. These futures resulted in *physical deliveries* of only about 12,000 tons per month, to which we can add the minor deals of various agents to get a figure for total deliveries that is almost trivial. (Of course, it would hardly make sense for African or South American producers to deliver their products into an LME warehouse when the final consumer was in Japan or France. There are eight LME warehouses and delivery places in the United Kingdom, in addition to warehouses in Rotterdam, Hamburg, and Antwerp.) As for the categories of buyers and sellers on the exchange who are interested in physical transactions, the largest group of buyers usually consists of merchants acting on the part of customers desiring to make marginal adjustments in their stocks; while the principal sellers include large producers selling small quantities, small producers selling excess production, and fabricators selling excess stock.

Furthermore, the trading of futures on the LME has to do with "hedging," or transferring price risk from buyers and sellers of physical commodities to "speculators." This function is important since commodity prices are extremely unstable, and later in this chapter the mechanics of hedging will be explained in detail. As an introduction to this topic, we can now take up the difference between the futures, forward, and spot (or cash) market.

The first thing to remember is that with the possible exception of a futures market, none of these markets need exist in the form of an organized exchange where all buyers and sellers or their agents can congregate. As pointed out earlier, the spot market is not necessarily a market, but rather an arrangement that concerns delivery in the immediate future. The forward market, on the other hand, involves "forward" sales, or the sale of an item that will be delivered in the future at some mutually agreed on fixed price, or perhaps a price related to the price or prices on a metal exchange at or around the time of delivery. In addition, the forward market involves physical delivery, which is specified in a forward contract.

In contrast, a futures market features physical delivery in only a minority of cases. Strictly speaking, a futures contract *is* a forward contract, but this market is so organized that sales or purchases of these contracts can be "offset," and future deliveries are unnecessary. The key element in this process is the presence of specualtors who buy or sell futures contracts with the intention of making a profit on the difference between sales and purchase price. This matter will be clarified later with the help of some simple numerical examples, but in brief the arrangement functions as follows: A producer sells a physical commodity for forward delivery at a price related to the price of the commodity on a metal exchange at the time of delivery. At the same time he *sells* a futures contract. Then at the time of delivery, he *buys* a futures contract, offsetting his previous sale. If the price of the commodity has fallen, he loses on the physical transaction; but if the price of futures contracts fall, as they should, then he has a compensating gain on the futures (or "paper") transaction. (The matter of how the price of futures contracts should move in relation to the price of the commodity will also be explained later.)

What has taken place in this example is that the seller of the physical commodity has turned over most of his price risk to the buyer of the futures contract, the speculator, who in this case might have thought that the price of the commodity was going to increase, in which case the price of the futures contract would also normally increase. Thus he would be able to sell this contract for a higher price than he paid for it. The difference between the producer and the speculator where this market is concerned is that the producer is primarily concerned with insurance and regards the buying and selling of futures contracts as being of secondary concern: in the long run, with extensive buying and/or selling of these contracts, he should break even. As for the speculator, his business is to make money on the difference between the price at which he buys the contract and that at which he sells it. Naturally, the speculator also sells contracts in the hope that he can later purchase similar contracts, and in the process make a profit (per contract) equal to the difference between the sales price of a contract and the purchasing price.

Speculation is also possible in physical items. If, for instance, the producer in the preceding example had produced a certain amount of metal and held it in inventory without hedging it, thinking that he could make a profit by selling it later at a higher price, then he could be called a speculator. The same applies to someone who buys a commodity on the spot market and holds it unhedged in hope of selling it later at a higher price. This position is called being *long* in the commodity. Another way of speculating is to be *short* in a commodity, which works as follows: A speculator agrees to sell a commodity in the future for a price above the price at which he "thinks" he can buy the commodity just before the date the purchaser is to receive the commodity. In other words, when the (forward) contract comes due, he buys on the spot market for a price that, if all goes as planned, is under the sales price on the contract. The term "short" simply means selling something that one does not own.

At various times exchange pricing has come under attack from producers, consumers, and interested outsiders. Some people say that the exchanges are responsible for the severe price swings that are typical for certain commodities. If we take the copper market, however, this contention overlooks the well-known fact that when the exchange was closed during the first part of the postwar period, all known quotations of copper prices showed a tendency to oscillate that was in no way different from what we have observed on the LME or COMEX over the past decade. Other complaints are that the exchanges are a tool of "Big Capitalism"; they can be infiltrated by the agents of international communism and prices rigged; they take the order of "high finance"; they favor the Third World; and so on. Some of this, to a fairly small degree, may be true, although I feel that many of the charges against the exchanges originating from the academic world can be attributed to a profound unconversance with the way these institutions work. On the other hand, it seems to be true that many producers of industrial raw materials are peeved with the exchanges because they feel that they raise the price level of these materials which not only attracts outsiders to these industries but also leads to "excessive" investment in new capacity.

Exchange Contracts for Copper

The details of the proposed (and probably forthcoming) exchange contracts for aluminum are not available at the present time; however, since they will almost certainly function exactly like the existing copper contracts, we can still review briefly the various contracts for copper on the LME and COMEX. Here it should be remembered that copper is, for all its importance, a typical industrial raw material, and an exchange contract for copper is analogous to those of any commodity traded on these exchanges.

At present there are three standard forms for copper traded on the LME:

1. Electrolytic or fire-refined, high-conductivity wirebars in standard sizes and weights.
2. Electrolytic copper cathodes, with a copper content of not less than 99.90%, or first-quality fire-refined ingot bars, with a copper content not below 99.7%.
3. Fire-refined ingot bars with a copper content not below 99.7%.

Anyone can buy or sell on the LME, and the only limitation is that the transaction must involve 25 L-T. The brand and place of delivery are chosen by the buyer, but it must be remembered that "place of delivery" means one of the LME warehouses. It is this provision that makes the LME of only limited interest as a physical market, although it does not reduce its attractiveness for hedging purposes, nor does it suffer a reduced efficiency as a market on which smaller amounts of metal are purchased for physical delivery. Each of the three types of refined copper traded has its own quotation for both cash and forward deals. The spot price is also referred to as a "settlement" price, while the "forward" quotation, which in practice is mostly a futures price, is a three-month price. In other words, if this contract is to be used as a futures contract, the offsetting arrangement must be made within three months of the time the contract is issued; or if it is to be used as a forward contract, then delivery must be made within the same period.

As for COMEX, only one standard contract form is available. The basic commodity is electrolytic copper in wirebars, slabs, billets, ingots, and ingot bars, of standard weights and sizes, with a copper content of not less than 99.90%. The standard unit for trading purposes is 50,000 lb. In addition to electrolytic copper, a number of other varieties of copper may be delivered at the option of the seller. These include fire-refined, high-conductivity copper, lake copper, and electrolytic copper cathodes. According to the regulations of the exchange, copper may be delivered from any warehouse in the United States that is licensed or designated by COMEX, but other warehouses are excluded. Deliveries must be to designated delivery points, and the period of forward trading must be within fourteen months. There are seven delivery months: January, March, May, July, September, October, and December. For more on this topic, and also for some of the effects of speculation in these markets, see Rowe (1965) and Banks (1976).

Hedging

This topic is important, and the reader who is intent upon mastering it in the shortest possible time needs first to grasp the following rule: Those wishing to insure against a fall in prices sell futures; those wishing to insure against a rise in prices buy futures. Individuals falling in the first category might be sellers of a commodity that will be delivered and paid for in the future while those in the latter category might be the purchasers of a commodity that will be received in the future and paid for at the same time. In both cases the relevant price is the spot price prevailing at the time of delivery. (Of course, price risk could also be eliminated here by using a forward contract with the price specified. But this is not always possible.)

Consider the following example. A producer sells a commodity for delivery 2 months in the future at the spot price prevailing on a certain commodity exchange on that day. Assume that 50 is the spot price of the commodity on the day of the sale, and 53 is the price of a 5-month futures contract. The producer contacts his broker and orders him to *sell* a futures contract for 53. Two months later he delivers his commodity, with the spot price on the exchange on that day being 45 and the price of a futures contract at 46. He thus gets 45 for his commodity and pays 46 for a futures contract. The contract he sold earlier has now been offset; and his deals can be summed up as follows:

+53	Sales of futures contract
+45	Sale of commodity
−46	Offsetting purchase of futures contract
+52	Realized on the sale of the commodity

The broker's commission should be subtracted from the +52 to get the net value of the sale. There are some other technicalities associated with the example that the reader should be aware of. On the day the sale was arranged, the spot price was 50 and the price of a futures contract 53. The difference between these is called the *basis,* which in the example equals 3. Notice also that the futures price is higher than the spot price, which is the normal situation and is called a *contango*. When, on the other hand, the spot price is higher than the futures price, which happens from time to time, we have *backwardation*. The insurance aspect of the hedge can now be noted. Had the producer held the profit unhedged, his revenue would have been 45. By hedging, he gets 52, from which the broker's fee is subtracted.

It should also be appreciated that for hedging to work satisfactorily, the basis must exhibit certain "regular" tendencies. There should be no frequent movements from contango to backwardation, nor should there be excessive movements in the value of the basis. This is what we meant earlier when we indicated that when the spot price increases, the futures price should also increase or, as in the case of the preceding example, when the spot price decreases, the futures price should follow it down. History seems to indicate that this is the usual arrangement, but it has happened that this pattern has been interrupted for short periods, causing hedgers some discomfort. Accordingly, the risk to hedgers is often called the *basis risk*.

By way of deepening our insight into this situation, let us change the preceding example so that the futures price on the date of arranging the sale was 51, and the price of a futures contract at the time of delivery is 52. The other prices—the spot price on the date of arranging the sale and the spot price at the time of delivery—will be left the same, 50 and 45, respectively. We then have a change in basis from 1 to 6, and the revenue from the sale, not including the broker's fee, is $51 + 45 - 52 = 44$ (sale of futures contract + sale of commodity - offsetting purchase of futures contract). The hedger has lost on this deal in the sense that, had he not hedged, he would have obtained 45 (or the spot price on the day of delivery). This loss may seem insignificant, but the reader should remember that it involves only one unit. Had the producer hedged 100,000 units, it would have been a serious matter. In practice, though, abnormal changes in the basis happen to be rare, and the regular hedger can be comforted by knowing that, statistically, he should break even in the long run.

As mentioned earlier, commodities are bought on the exchanges employing standard contracts. A standard contract on COMEX for copper is for 50,000 lb; thus if a deal involved 100,000 lb, two contracts would be bought or sold. There is also a time element associated with these contracts. In the first example, we mentioned that on the date of arranging the sale, the price of a 5-month contract was 53, and presumably this was the type of contract the producer sold. He delivered his commodity in 2 months at which time he bought an offsetting contract. Logically, the contract he bought was a 3-month contract—logically because the rule is that when one sells a futures contract, it involves a certain month, and when the offsetting purchase is made it must be made for the same month. The same is true if we begin the process by buying a contract: We buy for a certain month, and the offsetting sale concerns contracts for the same month. It is also possible to think in terms of a certain date: the *maturity* date. If someone has sold or bought a contract with a certain date of maturity, then before the date he must make the offsetting purchase or sale of a contract referring to that date.

The next example is a situation in which a buyer of copper arranges for delivery of 50,000 lb of copper 2 months in the future at the price prevailing on the exchange at that time. Assume the spot price on the day the copper is bought is 30¢ per pound, and the futures price 32¢ per pound. The basis, which is a contango, then stands at 2. The buyer then buys *one* contract, paying 32¢ per pound for 50,000 lb. If we assume that the spot price of copper increases to 40¢ on the day of delivery, with a futures price of 41, then the buyer makes the offsetting sale of one contract (or 50,000 lb) at 41¢ per pound and buys his copper at 40¢ per pound. His account now appear as follows:

-32	Purchase of futures contract
+41	Sale of futures contract (offsetting)
-40	Purchase of copper
-31	Price paid for copper

He pays 31¢ per pound for 50,000 lb of copper (to which he must add his brokerage fee). Notice what would have happened had he not bought and sold this futures contract: He would have paid 40¢ per pound for his copper.

It might be instructive to alter this example slightly. Assume a situation in which copper is bought for a *fixed* price of 40¢ for delivery several months in the future. On the same day the futures price of copper is 42¢, and so the buyer *sells* a contract for 42¢. On the day the copper passes to his hownership, its spot price is 30¢, and the price of a futures contract 31¢. The buyer then makes his offsetting purchase of a futures contract at 31¢. These transactions can be summarized as follows:

-40	Purchase of copper
+42	Sale of futures
-31	Purchase of future (offsetting)
-29	Price paid for copper (per pound)

Several points are important in the preceding example. Had the buyer not hedged, he would have bought copper at 40¢, while some of his fellow producers bought at 30¢ or 29¢, either by waiting to buy spot, or buying at 40¢ and hedging. There is also the matter of what the situation would have been had the buyer expected the price to rise dramatically. With this the case it could be argued that had the basis remained constant, it would have been better to buy at 40¢ and not hedge: Had the price risen to 65¢, and the basis stayed the same, then hedging would have meant selling a futures contract for 42¢ and making the offsetting purchase at 67¢. The cost of copper would then have been 40 + 67

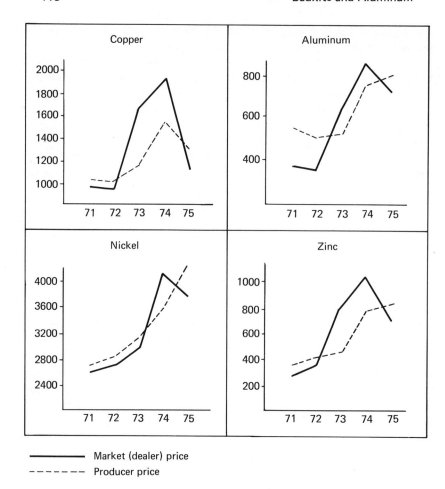

        ——————   Market (dealer) price
        ------   Producer price

Figure 6-1. Producer and Market Prices for Copper, Aluminum, Nickel, and Zinc (in Dollars per Ton)

– 42 = 65¢. This situation could conceivably come about, but in general we expect that the large price rise would have been preceded by a substantial rise in the price of futures contracts. Instead of selling a futures contract for 42, it might have been possible to call one for 55 or 60.

This section will be concluded with a few remarks about the commodity exchanges. There are no dealings on the COMEX in aluminum, but arrangements are definitely being made to introduce zinc on the COMEX, and nickel and aluminum contracts will soon make their appearance on the LME. At present, between 2 and 10% of the world's 14 million annual tonnes of aluminum metal output is sold on the free market, and thus at current prices turnover comes to less than $1.5 billion, of which about $10 million are the commissions of metal merchants. Plans now call for commissions on the LME to be lower than the 5 pounds/ton that is the market average, which simply means that the LME wants to make sure that as the world use of aluminum increases, and production increases in places like the Middle East and the Caribbean, the LME will serve as a focus for the trading of this metal. The LME will also trade in slightly inferior grades of aluminum (99.5% pure) in order to attract supplies from East Europe and secondary producers.

Figure 6-1 shows the relationship between the free market price and the producer price for four nonferrous metals. Although it cannot be discerned from these diagrams, the free market arrangement gave on the average higher prices during the period shown for these and other metals. Market prices were 8% higher for copper, 23% higher for zinc, and 7% higher for tin. Producer prices were 7% higher for aluminum and 2% for nickel. Statistically, there was no difference for lead. If, however, we go back to the period 1966–1970, we see that on the average the market price of copper was 45% higher than the producer price, and the market price of nickel 125% higher than the producer price.

Some Simple Analytics of Hedging and Speculating

The previous section began with an example involving someone selling an item to be delivered in the future at an unknown price. The proposition was that this person should buy a futures contract since, if the spot price of the item fell, when the contract was offset, it could be offset at a lower price, and thus the "loss" suffered due to the decrease in the spot price could be compensated for by the difference between the sales and purchase price of the futures contract. But what about someone selling forward an item at a *fixed* price. If this person becomes convinced shortly after the deal is closed that the price of the item is going to rise, then he or she might buy a futures contract. Again, assuming an unchanged basis, should the price actually increase, the purchaser of the contract would stand to profit. Notice the difficulty in categorizing the actor in this little drama because if this person's second thoughts about the possibility of a price increase were strong enough, regardless of the gain that he or she had already made on the sale, some people would argue that he or she should buy the contract. On the other hand, the owner of a futures ontract usually expects to sell a contract later, and regardless of his or her beliefs at the hour of purchasing the contract, something could happen to necessitate offsetting the contract at the lower price.

Some diagrammatic notation will be introduced at this point. What we have had up to now is a situation in which hedgers are both demanding and supplying futures contracts, depending on whether they are insuring against a price increase or decrease and speculators are demanding or supplying contracts on the basis of their beliefs about the price of these contracts in the future (which also says something about the beliefs of speculators concerning future spot prices since it has been claimed by some investigators that *the price of futures contracts* provides the best available estimate of *future spot prices*). Figure 6-2(a) shows the demand and supply for futures contracts by speculators. Note that as the price of contracts becomes greater than P_f^*, we have a *net* supply of contracts. The concept of "netness" must be emphasized here because even at very high prices, there may be some speculators who feel that the price of these contracts will go higher, and so they buy futures contracts. At this point the reader should note that if expectations were to change in such a manner as to make a sizable number of speculators believe that spot prices and, by extension, the price of futures contracts were going to rise, they would increase their demand and/or reduce their supply of futures contracts. This would cause the curve S^s-D^s to shift up.

As indicated earlier, hedgers interested in insuring against a price decrease

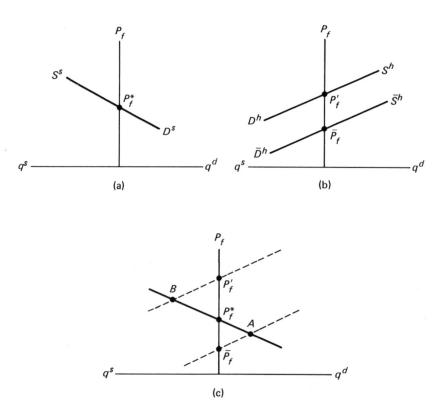

Figure 6-2. The Demand (D) and Supply (S) of Futures Contracts by Hedgers and Speculators as a Function of Their Price (P_f)

would increase their supply of futures contracts the higher the price of these contracts. Similarly, when the price of these contracts is very low, individuals desiring to insure against price increases have a very strong initiative to purchase futures contracts. Thus the D^h-S^h curve showing the behavior of hedgers runs up from left to right in figure 6-2(b).

Another typical curve showing the supply and demand of hedgers is $\bar{D}^h$-$\bar{S}^h$. This schedule, however, is based on a different set of expectations about future spot prices and/or the future price of futures contracts.

Next we put these results together, as shown in figure 6-2(c). Depending on which supply-demand curve we use for hedgers, we have equilibria at A and B. One important difference between these equilibria is that at A, hedgers are net suppliers of futures contracts, which would seem to correspond to statistical evidence from the real world. This may explain why some economists choose to ignore the possibility that there are some market participants who might find it advisable to insure against price rises. Thus at B hedgers as a whole are demanders of futures contracts.

We can close this section by mentioning that in July 1974 the prestigious *OECD Outlook* found it "remarkable" that speculators buying copper forward were systematically gaining, while those selling short were losing on their transactions. Given that the price of copper began rising at a rapid pace in the third quarter of 1972 and continued to escalate until the second quarter of 1974, it seems clear that traders buying for forward delivery at a price close to the existing spot price would stand a good chance of being able to sell at a higher price, while those selling short were forced to buy in a rising market. After the second quarter of 1974, however, this situation was generally reversed as copper prices began a long descent.

Commodity Prices and Inventories

We turn now to the link between commodity prices and inventories, but first a few remarks are necessary about long- and short-run prices. Long-run prices are determined by trend movements in supply and demand. If, for example, over a long period the supply of a raw material expands at a more rapid rate than demand, then we should expect an ineluctable downward pressure on price. This has been the case with copper until recently. In contrast, the demand for aluminum metal seems to be increasing faster than the building of new facilities for producing this metal, and given that the gestation period of investments in aluminum capacity is a minimum of three or four years, it has been claimed that we are moving toward a sellers' market in aluminum that may feature strong price increases. As things stand now, only a prolonged business downturn could alter this scenario.

On the other hand, when we examine short-term prices, we see peaks and troughs that are separated by weeks or months instead of years and that are to some extent independent of business activity. The explanation of these short-cycle price oscillations lies in the speculative tides of bullishness and bearishness fueled by fantasy, naiveté, or just plain bad judgment. Thus a relatively minor surge in demand might cause metal prices to rise for a few weeks, causing an influential group of market analysts to glimpse what they think is the beginning of a new golden age. Something like this happened with copper in the United States in April 1978 when the price temporarily jumped from 56¢ per pound to 63¢ per pound. Its failure to remain at that level can be attributed to the well-documented fact that unfounded enthusiasm is still no match for the realities of supply and demand, and these realities can be expected to surface as more attention is directed to deciphering various market signals.

As soon as a substantial number of market participants come to the conclusion that the underlying economic situation points toward excess supply, a downward movement in prices is only a step away. Bullish (or good) news is now discounted or disregarded, and any bearish news that might have been given short shrift when the market was rising is reevaluated. Speculators begin to sense a downswing and increase their sales of futures contracts, expecting to buy them back later at lower prices. This decreases the price of these instruments, and since some market participants take the price of futures contracts as a forecast of the spot price of various commodities *in the future,* the overall feeling of deterioration is reinforced.

As with the upswing, the downturn is brought to a halt by the filtering through of sufficient high-quality information to dispel the fog concealing what is actually happening on world markets. Much of this information, it appears, comes in the form of interpreting the significance for present and future prices of inventory sizes and movements. For example, at the present time the zinc, copper, and nickel industries are plagued by stock overhangs that tend to preclude any rapid upturn in price, even given temporary dislocations in production. Once again using copper as an example, we recently saw that although both Zaire and Zambia experienced difficulties in either producing or shipping copper, the price of the metal continued to fall. This could have been attributed only to excessive world inventories.

For the most part the economic literature dealing with inventories has been confined to the learned journals, where it can be conveniently ignored. But for our purposes some of the more elementary concepts associated with this literature must be examined since, with a little luck, they give us the possibility of completing the readers' introduction into the logic of short- and intermediate-run price movements.

To begin, if a producer's or a consumer's inventories are low, then each extra unit held in stock reduces the possibility that production somewhere will have to be scaled down because of an unforseen absence of some input. Remember that both producers and purchasers of industrial raw materials are bound by contractual obligations to their customers, and as a result inventories must be held even if there is an inverted relationship between the spot price of the commodities being stocked and all predictions of the future price. In other words, even if the expected money yield from holding and later selling a commodity does not cover such things as its storage cost, this negative aspect is counterbalanced by a *convenience yield* when the size of inventories is small relative to the amount of the commodity being used as current inputs in the production process. In this situation an effective price system must function in such a way as to ration existing stocks among existing and potential stockholders, which often calls for a departure from normality in the form of an inversion between present and future prices. This inversion is called *backwardation,* and it existed on the copper market only a few years ago in the wake of some misunderstandings about the future availability of copper metal.

The same type of reasoning makes it clear that if the amount of stock being held is large in relation to the amount being used as a current input in the production process, then there is little incentive to hold more. In these circumstances convenience yields are small, and stockholders require that the expected future price of the commodities being held is such as to cover storage, handling, insurance, and other charges—unless it happens to be possible to sell futures contracts at a premium, which in the stockholder's opinion would be sufficient to cover carrying costs. Otherwise these stocks are put on the market, driving down spot prices and widening the gap between present and expected future prices to an extent that holding the existing stock is justified. Figure 6-3 illustrates some of these notions for an unspecified commodity.

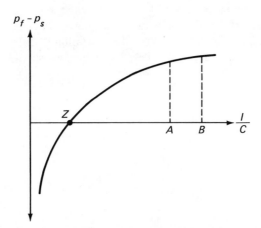

Figure 6-3. The Spread between the Expected Future and the Spot Price as a Function of the Ratio of Inventories to Consumption

In the diagram p_s is the spot price, and p_f is the *expected* future price of the commodity. A proxy for p_f might be the actual price of futures or forward contracts on one of the more important commodity exchanges. At Z we see the shift from backwardation, at low levels of the inventory/consumption (I/C) ratio for this commodity, to the normal condition called *contango* in which the expected future price exceeds the spot price. In relating this diagram to a real-world situation, it can be noted that some months ago copper inventories amounted to between 4 and 6 months of the annual consumption of that metal. Thus when fighting broke out in Shaba (Katanga) province in Zaire, the fall in world inventories corresponded to a movement from B to A in the figure. With expectations about future prices unchanged, this decrease in the I/C ratio was hardly enough to cause a ripple in the quoted prices of either the metal or the ore.

This section, like the previous, can be concluded with an example from the real world. In the London *Financial Times* for 6 February 1979, the distinguished commodity editor of that organ, John Edwards, pointed out that on the copper market at that time the gap between the cash price and three-month quotation was much smaller than it should be to carry supplies at the then existing interest rates. According to Mr. Edwards, this situation brought about suggestions that a shortage of supplies to the market could develop in spite of the large stocks being held at that time.

Now the total amount of copper being stored in the world is not determined by the desires of storers but rather by the physics of production and consumption. The level of world inventories is what it is, and the purpose of the market mechanism is to ensure that if storers are dissatisfied with this level at one structure of prices, that price structure is changed to one where storers *will* be satisfied. With no change in expectations, in the case given by Mr. Edwards, this would mean the selling of enough stocks to make the spot price of the commodity fall to the extent that the gap between the present and expected future price justifies the holding of existing stocks. My conclusion is therefore that, at least in the short run, no shortage of supplies "to the market" should be expected.

Appendix 6A
A Stock-Flow Model

We have already discussed some aspects of stock, flow, and stock-flow models. This appendix concerns a more profound elaboration of the stock-flow model, and in particular, derives an expression that says something about price movements in this type of model.

To begin, remember that a "flow" designates a quantity per unit of time at a point in time, while a "stock" is a quantity at a point in time. In the markets for many primary commodities, there is a demand for a stock to be held for speculative and precautionary purposes, and there is also a demand for these commodities for flow purposes—that is, as current inputs in the production process. We also have found it to be true that we can infer directly from actual markets that as the ratio of the inventory of an item to its current consumption increases, its price decreases. Chapter 5 suggested that the basic flow supply-demand model of the elementary textbooks did not make these things clear, and thus that chapter suggested an extension of that model. What we want to do here is consider a further extension—specifically, one that will include all those factors necessary for understanding price formation on the markets for nonfuel minerals.

We can begin by looking at a stock demand and stock supply curve. Stock supply is a datum. It is the amount of an item that exists at a given time and therefore is not a function of price or anything else. In contrast, stock demand is normally a function of present price, *ceteris paribus*. In other words, at any given time market participants have certain expectations about the future price of a commodity; and so if the present (or spot) price decreases, the spread between present and future price increases, thus making it more profitable to acquire a larger inventory. Eliminating the *ceteris paribus* clause, we should also expect a relationship between stock demand and, for example, the interest rate, expected future prices, and even flow demand.

Normally if the interest rate were to fall, the cost of financing a given amount of inventory would decrease, and thus the stock demand (which is the demand for inventory) would tend to increase. Similarly, if the present price were constant, an increase in the expected future price would cause the expected profit from each unit held in inventory to increase, and stock demand would therefore increase. Finally, it seems that the larger the amount of a commodity used for current production, the larger the amount held in inventory. Typical stock supply and demand curves are shown as $\overline{S}'$ and D in figure 6A-1(a).

The stock supply curve is simply designated $S = \overline{S}'$, while the equation for the stock demand curve is written $D = D(p_t; p_t^e, r_t)$. Here p_t is the price in period t, p_t^e is the price expected in some future period, where the time at which we consider the expectation is t; and r_t is the interest rate prevailing at time t. Obviously, many other variables could fit into the argument of this function. Flow supply and demand are shown in figure 6A-1(b) and are algebraically designated $s_t = s_t(p_t)$ and $d_t = d_t(p_t)$.

Notice the situation at A and A'. Here we have a full stationary equilibrium, with flow supply equal to flow demand and stock supply equal to stock demand. In each period q_0 is produced, which is consumed in the current production process. The stock of the item is constant at $\bar{S}'$, and the price is also constant and equal to $\bar{p}'$. Next assume that the expected future price increases. We show this by shifting the stock demand curve to the right: At every value of the present price, more of the item is demanded by profit-conscious stockholders. This increase in stock demand causes an increase in price, which leads to an augmentation in the stock of the commodity. This works in the following way.

With an increase in price, flow supply becomes larger than flow demand. As shown in the diagram, the price increase caused the present consumption of the commodity to fall, while profit-maximizing producers increased production. We now face a situation in which in every period the stock is increasing since as long as price is greater than $\bar{p}'$, we have an excess of production over present consumption. In the diagram price initially goes up *toward* p'', where the excess of production over consumption is shown by $s_2\text{-}d_2$. Given the model, this amount would be added to stocks during the first period *if* the price reached *and* stayed at p''; but remember that as various consumers obtain their stocks, the pressure on the market decreases and the price falls. For instance, by the end of the period, the price *might* have fallen to p''', in which case the addition to stocks would be some amount between $s_2\text{-}d_2$ and $s_3\text{-}d_3$. In figure 6A-1(a) the exact

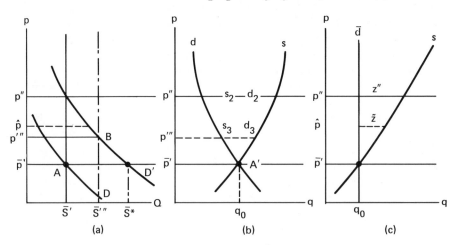

Figure 6A-1. Stock and Flow Demand and Supply Curves

amount is shown as being equal to $\bar{S}'''-\bar{S}'$. What the reader should pay atten-tion to here is the vagueness with which the value of price is delineated. The reason for the vagueness is that an extensive study of the price dynamics of this model, using linear supply and demand relations, has convinced me that it is impossible, a priori, to say exactly where the price will be after the initial dis-placement from equilibrium. Ideally, however, things will function more or less as they are being described here.

At the beginning of the next period we are at point B on the stock demand curve, and we go through the same sequence as before: excess flow supply, in-crease in stocks, and a decreasing price. Eventually, *if all goes well*—which means no oscillatory price movements—stocks will reach the desired level. For figure 6A-1(a) this means that the stock supply curve asymtotically, and smoothly, approaches the new equilibrium value $\bar{S}^*$. The price has again returned to the *full equilibrium* value $\bar{p}'$ where flow demand is once again equal to flow supply, and all production goes to current use. In other words, we have stationarity on all markets with flow demand equal to flow supply and stock demand equal to stock supply. Even though we do not have a full equilibrium at prices above $\bar{p}'$, we may have some *market equilibriums* because in some periods producers are producing exactly the amounts that buyers desire for current consumption and for additions to inventories.

Because of the time element, a mathematical exposition of the preceding situation can become quite messy. But for simplicity, suppose that instead of the falling flow demand curve shown in figure 6A-1(b), we have a verticle flow demand curve as shown in figure 6A-4(c). This type of flow demand curve is certainly appropriate for such items as aluminum and copper because of their low price elasticity of demand: in the short run we would not expect changes in price to influence the use of the item. We can then postulate two equilibrium situations. The first is a market equilibrium, and the relevant equations for this arrangement are given under column A in table 6-1. The second is also a market equilibrium, but at the same time it is a full equilibrium, with flow demand equal to flow supply, and stock demand equal to stock supply. This system is designated B in table 6A-1.

Note that in this table the last equation under column A is not given, it is simply designated. As it happens, this is probably the most important equation of the system, and later I will present what I consider to be a feasible candidate for this equation and solve it explicitly for price in terms of some of the system parameters. Apart from this exercise, mathematical manipulations with this model will be avoided since a verbal discussion of certain key issues is much more rewarding. To begin, the reader should look at equation (c) in table 6A-1, which is an equation for the investment demand i_t. To grasp what this is and how it is used, we can begin once more with a full equilibrium in figure 6A-1, with $S_t = D_t$. Now assume that an increase in the expected price causes the stock demand curve to shift to the right.

This means that initially, before the price increases, we have an excess stock

Table 6A-1
Simple Stock-Flow Model

A	B
(a) $s_t = s_t(p_t)$	(a') $s_t = s_t(p_t)$
(b)* $d_t = \bar{d} + i_t$	(b')*$d_t = \bar{d} = s_t$
(c) $i_t = k(D_t - S_t) = kZ_t$	(c') $D_t = D_t(p_t, p_t^e)$
(d) $D_t = D_t(p_t, p_t^e)$	(d') $S_t = D_t$
(e) $S_{t+1} = S_t + (s_t - d_t)$	
(f) $S_t = D_t$	
(g) Price equation	

*The flow demand curve is verticle, as in figure 6A-1(c). The total flow demand in a given period is then a constant $\bar{d}$ plus the investment demand. When we have a total equilibrium, as under column B, flow supply is equal to $\bar{d}$.

demand that, in figure 6A-1(a), is equal to $\bar{S}* - \bar{S}' = \bar{Z}$. The question that must be asked is if the price were to remain at $\bar{p}'$, would we want all of $\bar{Z}$ later or instead a portion of $\bar{Z}$ right away and the rest later on? Considering the first part of the query, is our *investment demand* equal to our *excess stock demand* in the initial period? If it is, then k in equation (c) is equal to unity. On the other hand, it could be that we want only a portion of $\bar{Z}$ in the initial period, in which case k would be somewhere between unity and zero, and the investment demand would be less than the excess stock demand.

Thus regardless of price, whenever we have excess stock demand, we must determine investment demand. It is investment demand rather than excess stock demand that is relevant in a given period. But once investment demand takes over, it becomes very difficult, if not impossible, to discuss this model with the help of simple diagrammatics. It is investment demand that determines how fast price rises and how far; and although algebraically this situation is not intractable, we may not be able to use the apparatus developed in figure 6A-1. The following example gives some idea of why this is so.

Assume an excess stock demand at the beginning of period t. Also assume that investment demand functions so that buyers do not want any of the commodity in period t. This is the same as saying that there will not be any investment demand in period t: The excess stock demand will begin to make itself felt only as an investment demand in later periods. Thus although we would show a shift in the stock demand curve in figure 6A-1(a), we could not justify an increase in price in period t. To repeat: We have an increase in the excess stock demand in period t, but no increase in price in that period so the arrangement shown in figure 6A-1(a) is inapplicable. What we are taking up now is a special situation, it is true, but it is certainly conceivable. Note, however, that this situation is not covered by equation (c) in table 6A-1 because with k constant and not equal to zero, there is a positive investment demand in every period that we have an excess stock demand.

We can continue in this vein with a few more examples since it is important that the reader get some insight into just how complicated the dynamics of these equations can be. Assume that $k = 1$ in equation (c), which means that investment demand is equal to excess stock demand. Now assume an increase in the expected price of a commodity at time t, leading to an instantaneous increase in the stock demand and excess stock demand. In the very small interval immediately after this change in the expected price, the present price is still $\bar{p}'$, and in figure 6A-1 we have an excess stock demand (and investment demand) of $\bar{S}^*$ - $\bar{S}'$. Then the effect of this investment demand hits the market, and the price goes up very rapidly toward p''.

Note that this investment (or excess stock) demand impinges first on existing stocks. In the very short run there is no possibility of increasing production, so those individuals who are certain that prices will increase buy from those who are less certain; those who think that the price will go very high buy from those who think that the price will increase by only a small amount, and so on and so forth. Observe also that in figure 6A-1(a) the price *could* eventually increase by enough to choke off excess stock demand. As shown, this price is equal to p''.

Assume now that producers go into action, taking the rapid increase in price as a signal that considerable production is warranted. But if they increase production by the amount z'' and attempt to sell it at price p'', they may not find any buyers. But if the price is slightly less than p'', say, $p'' - e$, then an amount can be sold equal to the difference between the stock demand and supply curve at price $p'' - e$. Continuing in this fashion, we can imagine an excess flow supply $\tilde{z}$ which, at price $\tilde{p}$, is equal to the excess stock demand at that price—and they are fortuitously brought together in the same period. This price, excess stock demand (which equals investment demand since $k = 1$), and excess flow demand constitute a market equilibrium, but not stationarity or a full equilibrium. Instead, the stock supply curve displaces to the right by the amount $\tilde{z}$, and we begin our sequence all over again. If things continue in this fashion, the price eventually approaches $\bar{p}'$, and stocks approach $\bar{S}^*$. In other words, we approach a full equilibrium, with all production being used as a current input. The reader should take note of the following, however. In this example we have assumed, implicitly, some very simple inventory behavior on the part of producers. For example, if at the end of a period, they saw that they had overproduced, they simply reduce the price until they sold all of that period's production. But since, as we have made clear earlier in this book, producers *are* in the habit of holding stocks, we might assume a type of behavior that had producers' unwilling to reduce prices below a certain level in the short run and thus putting their excess production into inventory. Making an assumption of this nature would mean, once again, that figure 6A-1—or indeed any diagram—would be of only limited value in tracing the course of events.

Before going to the algebra, we can consider a simple variation on the preceding theme. Remember that the initial price signal to producers, following the increase in stock demand, was p''; but at this price it was postulated that excess stock demand had been eliminated. But consider a situation in which it is possible to increase production very rapidly, and so when the price reaches the vicinity of p'', production is increased by z''. A mistake has thus been made: Based on an assumption that price will be p'', production has been raised by an amount that, had the price actually turned out to be p'' (and stayed at this level during the period in question), could have been sold without producers' taking a loss. But now, to sell all or part of z'' in the present period, the price will have to be reduced below p'', and with the given flow supply curve, producers will be taking a loss. Thus part of this production might be sold and the remainder put into inventory. But which fraction will receive which treatment? Even a simple measure such as this involves inventory costs as well as costs attributable to the unforeseen alterations in production that might have to take place as a result of unplanned inventory changes. As the reader might guess by now, this simple variation is actually quite intricate. It has led us into a situation of frustrated expectations, undesired producer stocks, and various disequilibria that probably can be treated only verbally case by case. Facile generalizations carried out with the aid of simple algebra and elegant diagrams are out of the question.

Now let us go to the price equation that was listed, but not spelled out, under (g) of table 6A-1. Essentially what we need here is a simple behavioral assumption, and the one I will use is that the change in price is directly proportional to the investment demand, increasing when this increases and decreasing when investment demand decreases. Ideally we would like to have a nonlinear relation here that could take into consideration such factors as excess capacity and faster rates of price increases as capacity becomes tight; but unfortunately I find the gap between wanting to derive such a relation and my ability to do so too much to bridge at the present time. The proposed equation is:

$$p_t = p_{t-1} + \lambda(D_t - S_t)$$

Note especially that λ can be a more complicated variable than it appears in this equation. For instance, since price movements should be a function of the investment demand, we might have $\lambda = \theta k$, in which as shown in table 6A-1, $k(D_t - S_t)$ is the investment demand in period t. We can now lag the previous expression to get:

$$p_{t-1} = p_{t-2} + \lambda(D_{t-1} - S_{t-1})$$

Subtracting, we get:

$$p_t - p_{t-1} = p_{t-1} - p_{t-2} + \lambda[(D_t - S_t) - (D_{t-1} - S_{t-1})]$$

or: $$p_t - 2p_{t-1} + p_{t-2} = \lambda[(D_t - S_t) - (D_{t-1} - S_{t-1})]$$

and: $$S_t = S_{t-1} + X_t$$

Thus substituting this identity into the previous expression:

$$p_t - 2p_{t-1} + p_{t-2} = \lambda[(D_t - D_{t-1}) + X_t]$$

Now we need an expression for $X_t = s_t - d_t$. As mentioned earlier, all supply and demand curves in this analysis will be taken as linear; thus we have $s_t = e + fp_t$ and $d_t = g + hp_t$. In addition, let us take $D_t = \alpha + \beta p_t$, which implies that $D_{t-1} = \alpha + \beta p_{t-1}$. Thus:

$$p_t - 2p_{t-1} + p_{t-2} = \lambda[\beta(p_t - p_{t-1}) + (e - g) + p_t(f - h)]$$

or: $$p_t[1 - \lambda(f - h + \beta)] + p_{t-1}(\lambda\beta - 2) + p_{t-2} = \lambda(e - g)$$

This is a simple second-order difference equation of the type $p_t + a_1 p_{t-1} + a_2 p_{t-2} = a$, and it can be easily solved. Without going into this very elementary matter, it is apparent that depending on the values of e, f, λ, and so on, we can get such information as oscillatory price movements since the roots of this equation could be complex. In other words, we do not have the smooth progression from disequilibrium to equilibrium discussed in connection with the previous diagram. Finally, our equilibrium (when it exists) is the situation in which we have $p_t = p_{t-1} = \cdots = \bar{p}$. Thus, from the previous equation:

$$\bar{p} = \frac{e - g}{f - h}$$

On the basis of the preceding equations, we see that this describes a *flow* equilibrium. But from the first equation in our exposition, we have $0 = \lambda(D_t - S_t)$, and since $\lambda \neq 0$, $D_t = S_t$. Thus we also have a stock equilibrium.

7 Ore Grades, Energy, and Investment

It can be said with some confidence that never in human history has more attention been paid the topic of the long-run supply of minerals than at the present time. Thanks to the oil price increases of 1973-1974, we were given an insight into the reactions of ordinary mortals when faced by a sharp decrease in the availability of a vital industrial raw material and source of power for the most popular consumer durable. For the most part, it was an edifying experience, although one that few people would like to see repeated in the near future.

Two schools of thought seem to have resulted from that period. The first is that mankind will be lucky to survive the next few hundred years, given the growing pressure placed on known and prospective mineral resources by increasing affluency in the developed world and accelerating population growth in the LDCs. In general, people taking this position are finding it more difficult to find platforms from which to air their views as we recede in time from that second week of October 1973 when OPEC signified that the citizens of certain countries in the industrial world would have to curtail their motoring habits whether they wanted to or not. Conversely, other people took the view that the intrinsic logic of market economies is such that suitable technology will always appear in such a manner as to permit advancing standards of living, even though there may be major changes, that is, increases, in the price of some industrial and consumer goods.

Some years ago I thought that I had resolved this issue, at least to my own satisfaction, in favor of the latter point of view, even going so far as to suggest that the escalating potential of technology might some day permit the resources of the moon to be classified as movable. Just now I am not so sure. Provided that sufficient energy can be made available, it seems possible that the mining of successively lower grades of various ores will permit Mr. and Ms. Consumer to continue their profligate exploitation of nature's bounty for many decades, or perhaps centuries, to come; however, I see no reason to believe that present-day consumer aspirations, based as they are on the tolerance, if not encouragement, of consumption without production, can avoid imposing a catastrophic debt on future generations—one they will have to pay regardless of their feelings in the matter.

It is true that predictions are always appearing that insist that one or another mineral is on the verge of exhaustion, and just as regularly new discoveries and technological advances emerge to save the day; but only a fool would be lulled into believing that it will always turn out like this. Petroleum and some metals

133

are almost certainly on their way into a short-supply situation; while assuring suitable access to others will, as Sir Kingsley Dunham (1974) concludes, result in environmental disturbances on a scale hitherto unknown.

We can now turn to a brief examination of the projected demand for some major metals over the coming twenty years. Most of the studies that I am familiar with seem to indicate that there will be a slowing down in the demand for minerals by the relatively affluent nations due to changes in the types of goods demanded; the effect of technological change on reducing the input required to produce a given output—which is true for the mining as well as the fabricating state—and substitution among raw materials in response to changes in relative prices, tastes, and technology. Reasoning of this type has played an important role in the work of Wilfred Malenbaum and his associates (1973), in which they stress that the knowledge, skill, and aspirations of humankind will work in such a way as to reduce the raw-material content of its consumption bundles.

The opinion here is that nobody in his or her right mind should believe that this is a plausible scenario for the near future. What is happening instead is that in developed countries, more individuals than ever are enjoying various goods and services whose real value is grotesquely out of proportion to the physical or mental effort they themselves have put into bringing these goods and services into existence. As far as I can tell, never in human history has it appeared so important, or even *essential*, to deplete at as fast a rate as possible the most easily exploitable deposits of ore and fossil fuels in order to provide large numbers of individuals making inessential or even counterproductive contributions to economic life with an advanced standard of living.

At the same time the Malenbaum report, to include its updated version, also envisages a world of more equitable income distribution, reduced unemployment and underemployment, high productivities in agriculture and the public service, an expanding world economy based on miracles of stabilization performed by fiscal and monetary policies in the developed countries (and perhaps some prodigies of policy coordination by that marvel of bureaucratic incompetence, the International Monetary Fund). There is also supposed to be a sizable increase in the volume of goods and services transferred from the developed to the less developed world. Dr. Herman Kahn has also preached this sermon.

With the exception of the stance on agriculture, most of this speculation should not be dignified by calling it a pipedream. There will, of course, be a growing movement of goods and services to the less developed countries; but although some of this—in the case of countries like Norway, Sweden, and Holland—will be due to the self-serving harangues of third-rate and overpaid academic scribblers and bureaucrats in the service of some aid or international organization, most of it will be in response to the exercise of monopoly power in the market for a vital economic input or inputs or because some countries have discovered how to inject cheap labor and substandard working conditions into some very highly technical activities. A certain insight into these matters can be gained from table 7-1 which gives some projections for the use of the most important nonfuel commodities in the year 2000.

Now cross-country input-output studies definitely show that the demand for mineral resources is proportional to industrial production or, for example, gross domestic product in the developed regions; while it seems to be true that in the less developed regions, the consumption of energy and natural resources increases more than proportionally with rising income. What this means is that deviations from trend growth in the demand for industrial raw materials (which is the significance of the difference between the historical and the various predicted growth rates) will most likely reflect on the trend rate of aggregate (or macroeconomic) growth. Granting that a great deal of flab can be eliminated from the economic organization of the affluent part of the world, the figures in table 7-1—to the extent that they can be taken seriously—still indicate that aggregate growth rates may decline by as much as 20% because of the absence, for one reason or another, of crucial raw-material inputs. To this can be added the

Table 7-1
Some Demand Projections for Three Important Raw Materials for the Noncentrally Planned Industrial Countries

	Demand (1974)	Average Demand (1971-1975)	Predicted Demand (2000)	Annual Growth (%)	Cumulative Demand (2000)
Iron ore (million tons of iron content)					
Malenbaum		431.9	919	2.1	
Leontief			1336		
USBM	513		1030	2.7	20,050
Historical	513		2450	6.2	33,710
Copper (million tons)					
Malenbaum		7.9[a]	16.9	2.75	
Leontief			22.5		
USBM	6.6		20.1	4.50	320
Historical	6.6		21.7	4.70	343
Primary aluminum (million tons)					
Malenbaum		12.28	36.52	4.30	
Leontief			75.50		
USBM	15.1		64.0	5.50	943
Historical	15.1		167	9.7	1734

Source: F.E. Banks, *Scarcity, Energy, and Economic Progress* (Lexington, Mass.: Lexington Books, D.C. Heath, 1977). Wassily Leontief (et al.), *The Future of the World Economy* (New York: Oxford University Press, 1977). Wilfred Malenbaum, "World Demand for Raw Materials in 1985 and 2000," summarized in *Engineering and Mining Journal,* January 1978. U.S. Bureau of Mines, *Mineral Trends and Forecasts,* Washington, D.C., October 1976.

cumulative negative effect on economic growth of declining productivity due to, for example, drug abuse, crime, and antisocial attitudes and sentiments that can be expected to follow the decline in material expectations that will be experenced in conjunction with steadily rising unemployment. On this last point let me note that with economic growth rates in the industrial world already too low to absorb increases in the number of people of working age into the active labor force, much less to reduce the existing pool of unemployed, it is impossible to avoid concluding that unless there are fundamental changes in the work ethic, the organization of work, and the system of incentives as well as a modification of the educational system from elementary school to postgraduate course, short-run economic reality should definitely contradict Malenbaum's euphoric prognostications before a great many years have passed.

Where the outlook for most, but not all, of the poorer countries is concerned, Malenbaum does not in his latest report see much chance for improvement, even though he seems to think they will be the recipients of an increasing amount of aid. The reason for his position, which with the exception of the last part, seems sound to me, is that any improvement requires, *first and foremost,* a reduction in the rate of growth of population, and unfortunately nobody knows how this is to be brought about. Of course, it must be admitted that there are a number of Swedish and Dutch economists who, reflecting an optimism promoted by years of well-paid attendance at United Nations and similar talkathons, are ready to lend their neurotic expertise to the position that impending technological breakthroughs, the redistribution of world income, and the intellectual fallout from some monster palaver in Geneva or New York could ensure an adequate supply of minerals and food for up to 20 billion souls. Several half-educated Norwegian "peace researchers" also aggressively support this position.

Ore Grades and the Price of Nonfuel Minerals

The work of Leontief, referred to previously, also suggests that there will be noticeable changes in the price of industrial raw minerals. In particular, his calculations indicate that, on the average, mineral resources will be almost three times as costly in relation to manufactured goods in the year 2000 as they were in 1970, with the big jump in prices coming between 1990 and the turn of the century.

As things now stand, this is going to be one of the most fertile areas of research for both natural scientists and economists because it is essential for decision makers in governments to know *now* something about the quantity and quality of natural resources they will have access to in a quarter of a century. Some work on this subject has been available for a decade or so, but for the most part it is uneven, and there seems to be a decided lack of consensus as to what we should expect in the way of future mineral prices. Here we can cite the historic study of Barnet and Morse (1963), historic in the sense that it raised these issues anew; as well as important studies by Nordhaus (1974), Phillips and Edwards (1976), and others who will be referred to later. What these scholars found was that the *real cost* of mineral output, measured in either labor units or capital *and* labor has declined since about 1890. This, of course, is not what we would expect a priori since although new discoveries have a tendency to add to known supplies of natural resources almost as fast as these are depleted by consumption and investment activities, the mineral content of the earth's crust is not unlimited. Thus as mineral stocks are exhausted and/or ore grades decline in the face of a historically rising demand, basic economic reasoning would lead us to expect a rising price for these minerals; but at least up to the beginning of the 1970s, things have not worked out this way. A pattern of fairly rapidly declining real prices also results if production costs are expressed in terms of labor (and resources), with capital costs being reduced to dated labor costs. Then we get the situation shown in figure 7-1.

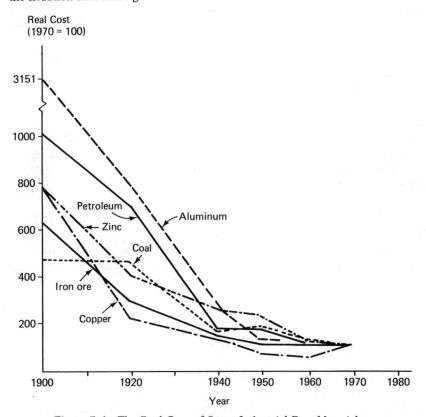

Figure 7-1. The Real Cost of Some Industrial Raw Materials

As reassuring as this may look, there are storm clouds over the horizon. Technological progress in the form of superior machinery and better educated labor and management has meant that by maintaining investment in the mining and processing industries at a high level, considerable progress has been made in reducing the value of primary inputs needed to obtain a unit of ore or metal. But questions have been raised by many of us as to how far this progress can go. The amount of rock that must be broken, handled, transported, and treated to obtain a unit of metal is steadily rising, which has necessitated the introduction of equipment that over time has tended to be progressively more energy intensive. The energy consideration was noted by Barnet and Morse, but in general was neglected until 1973 when it became apparent to all except a few perpetual optimists that energy costs, at least in the immediate future, were going to digress radically from the historical trend.

In addition, it could be argued that even if huge supplies of energy were to become available because of breakthroughs in, for example, fusion, there is a thermodynamic limit to the grade of ore that can be exploited. Both Phillips and Edwards, and D.A. Singer (1977) have made impressive analyses of this problem, and in conjunction with investigators like Lovering (1969) and Fischman and Landsberg (1972), their work suggests that a time might come when not only will the real cost of mineral output cease to fall but for some materials a steep increase in price might not lead to an increase in availability. Singer modifies this point of view somewhat by saying that although it may be unlikely that prices cannot be found at which at least a limited supply of all metals are available, these prices could be so high as to be tantamount to exhaustion.

A topic that is relevant to the present discussion is "Lasky's rule." Per se, this rule says that as the ore grade that can be mined and processed is lowered, we will realize huge increases in the reserves of most mineral resources. We would expect, of course, that as processing equipment becomes more efficient, a lot of the ore earlier discarded as waste would be reclassified as valuable, but a few optimists claim that if minable ore grades could be reduced by one half, reserves would go up by a factor of 10. The average ore grade of copper being mined today is 0.6 in the United States, and almost everyone considers the mining of ores with a grade of 0.3 to be a feasible proposition in the medium run, but as far as I know, there are very few expectations in industry circles of the United States' being other than a modest supplier of copper after the first quarter of the next century, at least if only land-based deposits of copper are considered.

There is also evidence from the microlevel suggesting that Lasky's rule is not universally applicable. For example, it may apply to portions of some medium-grade porphyry deposits such as the outside edge of the El Teniente copper mine in Chile, but according to Fuller (1976) it is not relevant to the deposit as a whole. This does not necessarily mean that as the ore grade of these deposits is lowered, larger supplies of copper or some other mineral cannot be extracted in the wake of the coming technological progress in ore refining; but it does mean that more and more material will have to be processed or, where processing equipment is concerned, very large amounts of energy may have to be used. The upshot of all this is that as the ore grade falls below a certain critical value, real costs may increase in such a manner as to necessitate extremely high metal prices.

Energy Problems and Ocean Resources

It now seems clear that due to the decline in ore grades, we are approaching an era when more energy input per unit of output will be needed for the mining and processing of many minerals. To get some idea of what the expression "more energy" means in this context, it should be appreciated that in the United States, production of the 12 largest energy consumers in the primary-materials industries requires 10% of the U.S. energy budget. It is also true that although copious amounts of aluminum are produced every year in that country, and this is the most energy intensive of the major metals, the total energy requirements of the iron and steel industry are three times that of aluminum.

Table 7-2 gives some idea of the *total* energy consumed in the production of several of the more important industrial raw materials in the United States as well as energy requirements per ton of metal produced (and energy requirements for secondary production). Although it may not be clear from the table, high-grade ores whose processing is characterized by relatively simple techniques generally have a low energy intensity. The only important deviation from this "rule" is to be found in the aluminum industry. The development of energy intensity over time is also of interest. As pointed out by Herbert Kellog (1977) in an important article, the average grade of the ore being mined in the United

Table 7-2
Energy Required for the Production of Primary and Secondary Metal
(*In 10^6 Btu's Per Ton of Material*)

	Primary	Secondary	Total U.S. Energy Consumption (10^{12} Btu's)
Aluminum	244	12.2	1100
Tin ingot	190	–	–
Nickel (cathodes)	144	15.0	–
Copper	112	18.0	221
Zinc	65	18.0	–
Lead	27	12.0	–
Steel[a]	25	13.0	3800

[a]Slabs; gray iron; castings.

States today is slightly over 0.6%; and it may be true that at the turn of the century this will be down to 0.3%. *If* this proves to be the case, roughly twice as much energy will be required per unit of copper concentrate recovered as is used today; and additional energy will be required to dispose of an amount of overburden that will be at least twice as much as that dealt with today.

Kellog's prescription for avoiding what he calls the "tyranny of falling ore grade" is to put more emphasis on recycling and the recovery of minerals from deep-sea nodules. If we look first at recycling, we see from table 7-2 that energy requirements for the most important metals are of a much lower order of magnitude than for primary output. Just what it will take to revitalize the secondary-processing industry is uncertain, however. The obvious suggestion is to increase incentives by providing a more favorable tax treatment for scrap collectors and processors; and also for manufacturers who build more recoverability into their products. Indeed, calculating just how much these tax incentives should be might provide a great deal of satisfaction for those analytically oriented economists who are looking for new subject matter with which to dazzle the editors, and perhaps even the readers, of our more prestigious learned journals.

Both the mechanics and economics of deep-sea mining are fairly complicated and can be only touched on here. This type of operation does circumvent the ore-grade dilemma in the form that we described it earlier; but as things stand, it will cost more per unit of output, in terms of direct and indirect energy inputs, than land-based mining. Deep-sea mining centers on the recovery of minerals from *nodules,* which are metallic objects whose size varies between that of a small pea and a cricket ball, and which can be found on the ocean floor at depths of from 900 to 6000 m. The average composition of these nodules is about 22.5% manganese, 1.25% nickel, 0.75% copper, 0.25% cobalt, and small amount of other metals.

An exact inventory of ocean resources has yet to be made. F.L. LaQue (1970) has estimated that the mining of only a small percent of the ocean floor would satisfy the current needs of the world for manganese and cobalt. These are two fairly important metals and could be even more important if they could be substituted for some of the other metals. A large part of the equipment that will be needed to lift these nodules has already been designed, and prototypes have been constructed. Already, apparently, there is some question as to just who has a right to call himself the owner of these underwater supplies—the firms or governments exploiting them or all humanity. Naturally, if it should turn out that large quantities of minerals could be recovered from the sea bed at some point in the near future, it would have important economic, and perhaps even, political implications. Countries like Japan would find their economic capabilities increased by enormous amounts.

There is evidence available indicating that all the oceans contain nodules, but the richest harvest is to be found in the east central Pacific. It may also be possible to mine the bottom of the oceans in a manner analogous to dry-land operations. After all, there is no reason why the mineralization of the bottom of the seas should in any way be different from that of the continental crusts. The problem cannot be attacked though without an entirely new mining technology—one that will probably be based on a new energy technology. Closer at hand may be the exploitation of marine placer deposits, located for the most part in shallow water near beaches and continential-shelf areas. The Red Sea has been denoted a possible target area for this type of deposit, and an intensive exploration program has been proposed. The metals that have been mentioned in this program include iron, zinc, copper, lead, silver, and gold.

As a caveat, it must be admitted that perhaps technology can come to the rescue again in the sense that it can substantially reduce the energy requirements needed to mine and process various commodities. It appears that the Aluminum Company of America may be on the verge of a major breakthrough in aluminum smelting technology. As mentioned earlier, there are a number of aluminum smelters in operation that use up to 12 kWh to produce a pound of aluminum, but the average is about 8, and in the most modern plants 6.5 kWh/lb of aluminum. But the Alcoa process, at least in its pilot form, is capable of reducing this to 4.5 kWh/lb of aluminum.

The new process is basically different from the traditional Hall-Héroult process discussed in the early chapters of this book. Both are electrolytic, but the Alcoa process uses aluminum chloride in granular form as its feedstock instead of alumina. Some questions have been raised, and for the most part left unanswered, about the capital costs of this new process, but Alcoa claims that for a smelter handling 135,000 tonnes per year, they would only be 10% higher than a Hall-Héroult installation. There is also good news on the environmental side, because the pots containing chlorine gas are sealed, and this new process does not emit fluorine. The key aspects of this project are secret, and apparently it cannot be brought to market for at least 5 and perhaps as many as 10 years; but just the news of its existence must cause a considerable stirring in Japan. On the basis of earlier discussion and the figures cited, if the Japanese can reduce the energy input to their aluminum industry by a third, or perhaps slightly less, it might well restore the competitiveness they have lost due to the increase in energy prices.

Before completing this discussion, it should be noted that just now, at least in the United States, the impetus to search for bauxite substitutes seems to be slowly waning. The reason for this is that at the present time (May 1979) there is no shortage of bauxite, and the danger of sharp price increases or supply cutoffs for this commodity is considered small. Besides, new sources of bauxite outside the IBA countries have reduced the world aluminum industry's dependence on the more volatile of the bauxite-producing countries. In the United States, only the U.S. Bureau of Mines appears to be moving ahead rapidly with a scheme to build a "miniplant" costing $50 million that will use other alumina bearing materials than bauxite as the input, although Reynolds metals has declared that it expects, within the framework of a 15-year research program, to come up with a viable process that can use domestic ores.

Alcoa, on the other hand, has apparently chosen to concentrate its developmental efforts on the energy saving process mentioned above and temporarily, forget about bauxite alternatives. Given the present day costs of turning alunites and other clays or silicates into aluminum, it is possible to understand the attitude of the U.S. aluminum industry. But it should be remembered that within 3 years of the formation of the IBA, taxes and royalties on IBA bauxite increased by 600%; and while bauxite now accounts for approximately 13.5% of the cost of producing aluminum, it is expected to rise to 15% by 1985. What it will *actually* rise to, nobody unfortunately knows, but a closing down of these research efforts will hardly encourage IBA to moderate their intentions to increase the price of bauxite at the fastest rate possible.

Investment and Pollution

This brief section opens with some remarks about "projected" investment requirements in the nonfuel minerals industries. These remarks center on table 7-3, the first part of which shows a substantial growth in the projected consumption of most nonfuel minerals in the noncentrally planned countries, despite the difficulties that have existed in the world economy over the last few years. One of the reasons for this fairly healthy growth of mineral input is that these figures are based on an assumed growth rate of income in the principal market-economy countries of 5%. The same assumption carries over to the projected investment columns at the right side of the table, which are also based on a 5% growth rate.

Since at present (1979), the projected aggregate growth rate in these countries appears to be settling somewhere between the 4.8% experienced in the period 1960-1970 (and predicted by the OECD for 1975-1985) and the 2.7% average growth rate recorded over the period 1970-1975, this assumed 5% would appear to be wide of the mark; but the reader can make a rough adjustment of the projected figures by simply taking his or her favorite growth rate (in percent), dividing it by 5, and multiplying this by the figures he or she is interested in. More sophisticated adjustments are of course possible, but they cannot be elaborated on here. At the same time it should be emphasized that regardless of the correct growth rate for the coming years, these figures postulate some interesting displacements in global production pattern for nonfuel minerals. For example, LDCs will produce less iron and more copper and aluminum.

As noted earlier, the suggested adjustment of the projected figures would also have to be applied to the last column. This is so because the investment projected in the last column is needed to bring capacity up to the point where it can provide the output shown in column 2. These investment figures, however, are at best approximations, and a brief canvasing of industry analysts and decision

Table 7-3
Some Production and Investment Data for the Nonfuel Minerals Industries

	World		LDCs		Investment Cost (1975) (U.S. Dollars Per Ton of Capacity)		Projected Total Investment Cost for Noncentrally Planned Economy	
	1975	1985	1975	1985	Mining	Processing	1978-1980	1981-1985
1. Copper	5,674	9,090	2,868	4,740	3,000-5,000	2,000	13.0	17.0
2. Lead	2,491	3,300	646	1,150	-	790-1,400[a]	0.87	0.82
3. Zinc	4,410	7,400	1,092	2,100	-	800-1,400	2.50	3.75
4. Bauxite	68,930	155,000	37,415	93,000	85	-	-	-
5. Alumina	21,489	47,000	4,957	11,500	-	510-750[b]	-	-
6. Aluminum	9,903	23,000	842	4,000	-	1,900-2,800[b]	13.40[x]	25.40
7. Iron ore	593	990	243	370	65-115	-	32.60	47.30
8. Phosphate rock	76,751	137,214	29,742	15,279	50	-	1.55	1.20
9. Tin	167	211	150	190	10,000-15,000[c]	-	0.45	0.11
10. Nickel	560	1,156	204	420	11,000-31,000[d]	-	7.40[y]	9.05[y]
11. Manganese ore	6,000	10,500	3,700	6,500	235-400	-	1.20	1.40
Column	(1)	(2)	(3)	(4)	(5)	(6)	(7)	(8)

Source: F.E. Banks, Scarcity, Energy, and Economic Progress, Lexington, Mass.; Lexington Books, D.C. Heath and Company, 1977. World Bank and United Nations Documents

Units: 1. Production in thousands of metric tons, with the exception of iron ore, which is in millions of metric tons.
2. Total projected investment in billions of 1975 U.S. dollars.
3. Copper, tin, zinc, nickel, and manganese ore in metal content.

[a]Mining-refining.
[b]First figure for integrated, second figure for nonintegrated facilities.
[c]15,000 dredging; 10,000 gravel pump installations.
[d]11,000-22,000 for nickel-oxide ore.
19,500-31,000 for nickel-laterite ore.
[x]Bauxite-alumina-aluminum.
[y]Includes centrally planned economies.

makers reveals a wide range of guesses about industry capacity in the middle of the 1980s. The investment surveys of the *Engineering and Mining Journal* indicate expected capital expenditures in the nonfuel mineral sector of $8-9 billion per year until 1985; while industry spokesmen seem to be thinking in terms of $14-15 billion per year. There are several ways of looking at his discrepancy. The industry estimate, as far as I can tell, says something about required investments if output continues to develop at the historical rate. On the other hand, if the lower figure turns out to be true, it means that there will be a significantly lower level of economic growth in the western market-economy countries, or a shortage of raw-material inputs and a much higher price for those that are available, or both.

Where the cost of additional capital is concerned, the reader should remember that we live in a period of high inflation and that equipment costs seem to be increasing by a minimum of 8-10% a year. Moreover, these costs vary widely and depend on such factors as where the particular installation is to be located, by-products, ore grades, annual output, and whether the installation is on the surface or underground. Significant expenditures are also necessary for infrastructure, and this is not just true for LDCs. Between 1960 and 1970 infrastructure costs for 11 projects in western Australia came to $515 million of a total investment cost of $800 million.

It appears that the trend is toward larger projects designed to take advantage of the increasing returns to scale realizeable in many activities of the mineral industries. Projects costing more than a billion dollars have long ceased to be uncommon. Capital-output ratios also seem to be increasing and range up to 4:1 for some of the newer installations, with investments of up to $250,000 needed to provide one job for a production worker. Financing has also become a complicated matter in comparison to the era when the mining industry consisted of relatively small installations and was largely self-financed.

Some new projects are financed by up to 80% borrowing, although in principle an attempt is usually made by firms in this sector to maintain a conservation capital structure. For instance, in 1969 the U.S. mining industry raised $1350 million by the sale of equities, and only $350 million by borrowing (while for the manufacturing sector in the United States, 70% of the securities issued were bonds). Whether this arrangement can prevail remains to be seen. At present the stock market is the worst possible investment medium, while the supply of loan capital, thanks to the Euromarket, is buoyant.

Next we go to pollution and pollution abatement and how they influence the production of nonfuel minerals. As is well known, mining, smelting, and refining are capable of generating copious amounts of pollution. For the most part the technology exists to suppress a large part of this pollution, but in doing so, considerable expenses must be incurred in the sense that a portion of the investment cost of a given installation is not available for producing conventional output. Instead, its function is to make sure that conventional output does not contain too many objectionable pollutants. In the United States, pollution-control legislation probably reduces overall or aggregate economic growth by 0.1–0.2 percentage points per year and could range up to 0.5% per year if all antipollution goals were realized.

For the nonfuel minerals industries the issue reduces to one of lower rates of return on invested capital. Beyond a doubt, given the uncertainty hanging over this industry as to the stringency of future regulations as well as the intentions of the authorities to police them, much larger risks are now associated with making many investments. In addition, having to pay attention to increasingly complex regulations, some of which reveal a profound misunderstanding of both economics and engineering, will undoubtedly introduce some inefficiencies into the resource-allocation process and cause the exclusion of considerable economic activity that might have been beneficial to all concerned.

Even so, I personally am convinced that pollution hazards, to include those from the mining and processing sectors, are so dangerous as to make their reduction worth the fraction of a percent in economic growth they now cost or will cost. But at the same time, everything else remaining the same, I prefer the subsidization of pollution control to the penalization of emissions, at least at the present time. In the long run a subsidy of this type would largely be self-financing, due to a reduction of health-care expenses—again assuming that all else remained the same; but in the short run some type of financing measure would have to be designed. One possibility is a pollution tax, with the tax proportional to the pollution generated by the item in its production or its use, and passed on in its entirety to the consumer. This person would then have a direct economic incentive for avoiding those products responsible for large amounts of pollution, the assumption here being of course that the subsidy would be so crafted as not to compensate a producer whose revenues declined because the pollution taxes on his output led to consumer rejection or who was too extravagant with his subsidy.

We can conclude this chapter by saying a few words about the aluminum industry and its pollution problems. The metal aluminum is not dangerous unless taken internally in fairly large doses, but some difficulties arise both with the production of alumina and the smelting of aluminum. With alumina the key issue is disposing of the red mud that is corrosive to the human skin and that can produce considerable damage to the ecology of water systems if dumped at sea or in swamps (as was once the case in Australia).

But it is the smelting of aluminum that has caught the main part of environmentalists' attention. It has been demonstrated in Sweden that the emission of fluoride impurities from refineries in sufficiently large quantities can destroy plant life in the vicinity of these installations, and by extension still larger concentrations are probably dangerous to humans. As a result, over much of the world quantitative limitations, which vary from country to country, are placed on these emissions. It also happens that employees of alumina- and aluminum-producing installations that are not safety conscious risk coming into contact with substances such as red mud dust, various flourides, and tars that can cause lung damage. Substantial investments may be necessary to eliminate these risks, and although it is difficult to get figures, it appeared that in Norway in 1973 a halving of flouride emissions added 8-10% to the cost of a medium-sized installation. Unfortunately, it seems to be true that above a 50% level of elimination, costs show a tendency to rise at a very fast rate.

Appendix 7A
Depletion Allowances
and Taxes

After considerable reflection, I have decided to postpone my plan to do a detailed survey of mineral taxation and depletion practices until a later publication; but given that most readers have some knowledge of what a tax is, I have allowed myself to be convinced that a few general remarks on depletion are necessary in this book. These remarks are general in the sense that they *could* apply anywhere, but the rules referred to below have their origin in the United States.

Depletion provides for the recovery of money invested in a wasting asset in the same manner that depreciation does for a capital asset. The way this is done in most countries is to subtract an annual depletion allowance from gross income when computing taxable income. In the United States a formula has been developed for determining depletion in any given year. This is known as "depletion by cost" and is:

$$\text{Depletion for year} = \text{depletion rate} \cdot \text{unit sold}$$

$$= \frac{\text{cost of property}}{\text{total units in property}} \cdot \text{units sold}$$

Another means for getting the annual depletion for tax purposes for oil, gas, and mineral properties is to take a percentage of gross income, providing that the amount allowed for depletion is less than 50% of the taxable income of the property before depletion. For the most part percentage depletion will be used if it results in less tax than depletion based on cost. In the United States some typical depletion allowances are:

Oil, gas wells, sulphur, uranium, bauxite, cobalt, lead, manganese, nickel, tin, tungsten, vanadium, zinc	22%
Gold, silver, copper, iron ore	15%
Coal, lignite	10%

We now construct an example in which we see the working of these two rules. Let us take a mining property that was purchased for $2500. The property contains 50,000 units of ore that will be extracted in 2 years at a rate of 25,000 units per year. The revenue per year will be taken as $12,000. Total operating expenses are divided into salaries and wages, and other inputs. The first of these is taken as $5000, and the second $1000. There is also some equipment associated with the mine, and the yearly depreciation on it has been calculated at $250. Now we can calculate depletion.

On the basis of the preceding formula, the depletion rate is 2500/50,000 = 0.05. The depletion allowance for the first year is thus 0.05 · 25,000 = $1250, and we have an identical calculation for the second year. Thus having computed depletion by cost, we can use it to obtain one value of the taxable income:

Gross income		12,000
Wages, salaries, other inputs	−6,000	
Depreciation	−250	
Depletion	−1,250	
	−7,500	−7,500
Taxable income		+4,500

With a tax rate of 20%, taxes are $900. However, if we use a percentage of gross income to obtain our annual depletion and, for the purpose of this exercise, assume that the percentage is 22, we get for the first year:

Gross income		12,000
Wages, salaries, other inputs	−6,000	
Depreciation	−250	
	−6,250	−6250
Taxable Income before depletion		+5,750
Depletion allowance = 0.22 · 12,000		−2,640
(This is 22% of gross income, and since it is less than 50% of taxable income before depletion, this method of depletion may be used.)		
Taxable income		+3,110

If the tax rate is 20%, taxes are $622. In this example, the 22% depletion rule is more favorable for tax purposes.

8

A Survey of the World
Nonfuel Mineral Economy

The first sign for most of us that all was not well on the natural-resource front appeared with the publication of Meadows and Meadows' *The Limits of Growth* (1972) and its portrayal of a world in which energy, minerals, and even fresh air would be in short supply. This report was followed by other doomsday epistles, and finally the oil price increases of 1973–1974 provided a nice example of history catching up with theory, at least to a limited degree.

Unfortunately, however, this example was not driven home with sufficient emphasis, except perhaps to the 8 or 10 million "extra" unemployed whose involuntary absence from today's work force stems directly from that crisis, as well as a few million others who have been forced to take or stay in jobs they do not want. As late as the autumn of 1974, a gentleman associated with the Swedish publication *Ekonomisk Debatt* could declare that the doomsday prophets based their work on theories that *we* find doubtful and strange, where "we" in this case represents academic mediocrity in its most grotesque form.

The idea of raw-material shortages appeared well before the dawn of modern economic theory and, in addition, was well documented. It simply happens that until recently most of us considered the topic too esoteric to give it any attention. The great eighteenth-century economists Malthus and Ricardo were among the first to mull over these topics, albeit superficially, and from the point of view of occasional shortages of arable land in relation to population. But the idea of a resource becoming extinct, in the fashion of the dinosaur or dodo bird, did not play much of a role in their work. Their position, translated to current usage, was that the substitution of one raw material for another and the force of technical change would serve to keep the concept of raw-material exhaustion out of the parlor of economics.

This thinking, of course, is both right and wrong. Technological progress, and substitution in both consumption and production guided by the price system, offer a powerful antidote to shortages and the misallocation of resources. Unfortunately, however, they cannot perform their miracles in zero time; and we have learned—or should have learned—from recent events in Vietnam and Iran that history does not always take the inertia of decision makers and their experts into consideration when it decides to change governments, fill the streets with broken glass, or present the more excitable members of the general populace the opportunity to redress real or imagined grievances.

Somewhat later the English economist Stanley Jevons made the possible exhaustion of coal the subject of a book, and the same fuel was taken up in a

well-known essay by the Swedish economist David Davidson. The beginning of
the 1900s saw the formation of a conservationist movement in the United States,
and even President Theodore Roosevelt found time to abstain from his posturing
and imperialist daydreams long enough to express a certain sympathy for the
goals of the conservationists. In 1907 on the first day of a conference of state
governors, President Roosevelt told his distinguished audience that many resources such as coal, oil, gas, and iron ore were certain to be exhausted some
day, and as a result it was necessary to see that they were wisely used.

During the Korean War, when it appeared that the United States-Soviet
Union confrontation was liable to be both long and vicious, President Truman
appointed the so-called Paley Commission to look into the scope of the United
States' inventory of natural resources. Their opinion, published with considerable fanfare, was that there was no risk of the exhaustion of any of America's
vital raw materials, either then or in the foreseeable future; and a tacit ratifica-
tion of this opinion was submitted by Cooper and Lawrence (1975) in their brief
perusal of the Paley investigation. This view of nature's benevolence seems to
prevail now on a worldwide basis, at least among those in a position to make
their convictions known to the rest of us. The eminent physical scientists Harold
Goeller and Alvin Weinberg have forecast a new golden age of growth and prosperity in about 200 years, based on what they say are almost inexhaustible
supplies of iron ore and aluminum-bearing clays; while a group of amateur
Scandinavian economists would like us to believe that the earth's crust is sufficiently chock full of minerals and other good things to permit the organizing of
a nice bourgeois standard of living for any number of inhabitants of our small
planet, provided that the citizens of the so-called rich countries adhere to the
preposterous verbiage emanating from assorted United Nations talk shops and
divert a share of their worldly goods to their less fortunate brethren in the
Third World.

My position is that a large part of this optimism is unjustified. Neither
Goeller and Weinberg, nor anybody else, has a usable prescription for traversing
the two centuries or so until their iron and aluminum paradise is in operation;
while the futility of direct transfers to the LDCs is obvious to everyone who
does not have something to gain personally by them. Moreover, if the population
growth of most of the Third World were to achieve and maintain a level just two
thirds of that of Mexico or Algeria, which strikes me as being quite possible,
then no conceivable rate of new discoveries of natural resources nor acceleration
in the rate of technical progress will prevent a series of disasters of the Sahel or
Bangla Desh varieties from eventually descending on a large part of the under-
developed world, probably affecting the industrial countries in one detrimental
sense or another.

Furthermore, although the global production of raw materials and energy is
still actually increasing, the 17 million unemployed of Western Europe, North
America, and Japan should provide a graphic illustration of the abject inability

of most governments, even in a situation of relative plenty, to "organize" the solution of a problem that threatens to erode, if not annihilate, the social fabric of the industrial world. Under the circumstances, some question must be raised as to both the sanity and honesty of the well-paid but unproductive young pseudoresearchers in countries like Sweden, Norway, and Holland who insist that now is the time for the working populations of the "rich countries" to open their hearts and empty their pockets in response to the poignant appeals of United Nations rhetoricians and their overpaid but underworked advisors and consultants. In fact, the growing hostility in Switzerland, Germany, France, Denmark, and Sweden against immigrants doing socially *essential* work should give some idea of the enthusiasm with which a summons to lower expectations and reduce consumption for the benefit of anonymous multitudes half a globe away would be received if, as is always possible at the present time, some dramatic event were to suggest a forthcoming shortage of natural resources.

The General Background

We can begin our main discussion with a few definitions. The first term of interest to us is *natural resource,* by which we mean a substance in the ground, water, and air that, in principle, is valuable to mankind. *Resources* are the part of natural resources that are quantifiable and available (at least in a conceptual sense) and which in addition can be used in existing industrial or consumption activities. *Reserves* are then the part of resources that not only can be used but that *will* be used given the technology of the period as well as the prevailing structure of prices and costs. On the other hand, reserves are generally considered to be the most important of these categories, and thus, going in the other direction, resources are often classified as known or hypothetical reserves that are not exploitable given the existing technology. Finally, raw materials are those reserves (and *mixtures* of reserves) that are usable in one form or another in industrial or other processes.

If we examine the crust of the earth, which is the part that interests us in that it is where most mining operations take place, we find large amounts of carbon and aluminum in addition to enough copper, zinc, lead, nickel, tin, and a few others to make these items the mineral basis of modern industrial life. Aluminum-bearing materials, for example, form a twelfth of the earth's crust, which means that anyone who has picked up a handful of soil anywhere has probably come into contact with an existing or potential basic ingredient of aluminum. Aluminum can also be found in oil shale and thus someday could become a by-product of petroleum. Another of its more recondite forms is as a component of the fly ash that is emitted by many coal-burning furnaces.

If we go deeper into the earth's crust we find increasing amount of magnesium and, continuing toward the core of the earth, a very high concentration of iron and nickel. Note that at present mining operations have in general not penetrated farther down than a few thousand meters, and to reach greater depths will require a great deal of highly sophisticated scientific and engineering work, in particular the design of instruments and equipment that can be used under extreme pressure and heat.

Similarly, an important frontier for mineral exploration is the floor of the oceans. Remember that of the 510 million km^2 that constitute the surface area of the earth, water claims 361 million km^2. (The rest is divided as follows: 65 million km^2 to the artic, Greenland, various tundra and deserts; 52 million to inhabitable but uncultivatible earth; which leaves 32 million km^2 of cultivatible land, of which 14 million is actually cultivated). As pointed out in an earlier chapter, if the manganese, nickel, cobalt, copper, tin, and so on that can be found at the bottom of the seas in so-called *nodules* could be brought to the surface, the global supply of these materials would be increased by very large amounts. A similar augmentation of mineral supplies would be possible if seawater could be "mined." Every cubic kilometer of seawater contains approximately 37.5 million tons of solids in solution or suspension, and in this bounty we find such minerals as iron, gold, copper, silver, nickel, and, especially, uranium. Getting at these materials, however, will require a technology that is radically different from any available at the present time, particularly in regard to its energy requirements.

Then too it should be noted that 1 km^3 of granite or shale contains 230 million tons of aluminum, 130 million tons of iron, 260,000 tons of tin, 7000 tons of uranium, 13,000 tons of gold, and so on, albeit in very lean concentrations. As a result the scientific work that must be carried out to gain access to these particular supplies is formidable, and several distinguished geologists have pointedly insisted that the average rock will never be mined. Just the opposite is implied by library-bound experts on depletable resources like Wilfred Beckerman when they use figures in the tens, or even hundreds of thousands of years when presenting their estimates of the total availability of some indispensable raw materials. Resolving this incompatibility of opinions is something that cannot be attempted within the confines of this modest survey, but as far as most rational economists are concerned, there are two key issues that must be taken into consideration.

The first is that the exponential growth of consumption of some depletable resources is capable of reducing the length of life of even astronomic supplies of these resources to a few hundred years. Professor Beckerman apparently overlooked this simple truth in his humble theorizing, just as he and people like him continue to disregard the fueling of this growth by population increases in the LDCs and the intensification of the consumer ethic in step with the dilution of the work ethic in most of the industrial countries. Equally important, trendy politicians and academics have lately begun to consider it their duty to erect every conceivable barrier to the creation of an educational system capable of training individuals who can produce these new resources and make the most efficient possible use of them once they are available. Some of our influential economists, of course, make it their business to avoid this matter, because in many universities the training of students takes a very poor second place to the droll courting of attention, popularity contests, and econometric escapades that constitute a large part of contemporary economic research.

The Mining-Processing Cycle

The next step in our exposition is to take up the mining-to-use cycle that the typical raw material experiences. As figure 8–1 depicts, the cycle begins with the

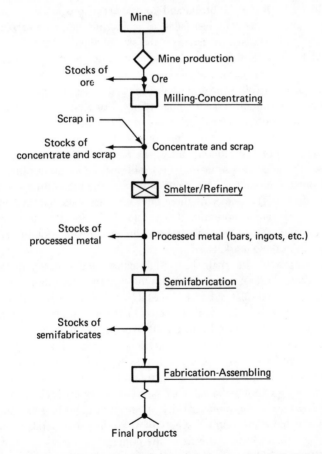

Figure 8–1. Mining-Refining-Final-Use Cycle for a Typical Industrial Raw Material

mining or extraction of an ore. A small amount of this ore may go into inventory, and the rest will be subjected to some preliminary processing that features such operations as milling, drying, and concentrating. Once completed, some of this "concentrate" will be stocked, but the majority will continue to the smelter and/or refinery. Smelting and refining can be two distinct steps, as with copper; or in some cases one of these is dropped. With aluminum, for example, we generally talk only of smelting, but some people use the expression "refining" in regard to the principal processing stage for this mineral.

We now have a product that is almost a pure metal. It is this refined (or in some cases smelted) product that is the center of interest for economists since because of its homogeneity, it can be sold on commodity exchanges or via trade journals without the buyer having to examine every item to make sure that it does not deviate too widely from his specification. Considerable stocks are usually held by both the producers and consumers of processed metals, and in the case of some of these metals (such as lead, zinc, and to a certain extent, copper), very good statistics exist as to the size of these stocks. It is also true that a verifiable relationship exists between the price of many metals and the size of these stocks. This phenomenon is explained in chapters 5 and 6, but roughly we have the following situation. Stocks guard both the producers and consumers of a commodity against uncertainty. For example, when stocks are low but consumption high, they feel exposed to the vicissitudes of fortune because a chance exists that they may not be able to fulfill their various commitments to their customers. Consequently, their attempt to maintain or rebuild stocks causes the price to rise in that they will not reduce stocks further or abstain from putting more of the current output of the metal into stocks unless paid well to do so. (The reader of these lines who has some background in economics should therefore note the the simple *flow* model of the elementary textbook in which stocks do not appear is an inadequate tool for explaining price formation in these markets.)

The next step in the chain is semifabrication where bars and ingots are turned into sheets, tubes, and various structural shapes. Sizable inventories are usually held of semifabricates; and it appears that this is a highly profitable portion of the cycle. For instance, if we consider the aluminum-processing cycle, bauxite mining seems to have been, on the average, the most profitable operation in this cycle, but at times semifabrication was even more lucrative. These semifabricates then serve as inputs to manufacturers of final products such as tennis rackets and automobiles. As an introduction to the next portion of our discussion, table 8-1 shows the value of various raw materials at the mining and metal stage as well as some other data that the reader should find useful.

One of the more interesting aspects of the nonfuel minerals industries is the manner in which different stages of the mining-final-use cycle tend to take place in different parts of the world. Many Third World countries, and even the Republic of South Africa, export a large part of their mineral output un-

Table 8-1
Production Data, Estimated Average Growth Rate to Year 2000, and Most Important
Uses for Some of the Most Important Nonfuel Minerals (1973-1974)

	Value Mining Stage	Value Metal Stage	Mine Production (000 Tons Metal Con.)	Expected Growth Rate to Year 2000[d]	Most Important Uses as Percentage of Total Use
Iron	7.2	48.60	490,000	3.06	Transportation (30), Building and Construction (28), Machinery (20)
Manganese	0.42	2.16	9,100	2.86	Transport (21), Building and Construction (21), Machinery (22)
Chromium	0.15	0.81	2,200	2.73	Building and Construction (22), Transport (17), Machinery (15)
Molybdenum	0.30	0.36	89	4.43	Transport (28), Oil Production (25), Machinery (22)
Wolfram	0.28	0.27	39	3.13	Machinery (74)
Minor ferrometals[a]	0.24	0.45	-	-	-
Copper	10.24	12.87	7,500	4.28	Electrical and Telephone (51) Building and Construction (17)
Zinc	1.53	3.06	5,800	3.24	Building and Construction (33) Transport (23)
Lead	1.23	1.53	3,500	2.27	Transport (52) Chemicals (11)
Tin	0.93	1.17	220	1.33	Packaging (35) Electrical and Telephone (18)
Aluminum	0.57	9.54	16,000	5.66	Building and Construction (22) Transport (22) Electrical and Telephone (14)
Magnesium	-	0.27	-	-	
Gold[b]	4.20	4.32	1.30	-	Decorative (60), Electrical and Electronics (30)
Platinum (group)	0.60	0.63	0.15	-	Chemicals (42), Oil Refining (18), Electrical and Electronic (17)
Silver[b]	0.78	0.81	9.40	-	Decorative (28), Photographic (25), Electronic (24)
Quicksilver	0.60	0.10	8.6	-	Chemicals (34), Electricity (28)
Others[c]	1.50	0.27	-	-	-
Total	30.00	90.00 (Billions of dollars)			
Estimated GNP growth rate				4.175	

Source: USBM and World Bank documents. Various U.S. government documents. F.E. Banks, Scarcity, Energy, and Economic Progress (Lexington, Mass.: Lexington Books, D.C. Heath, 1977).

Note: All values in billions of dollars.

[a]Cobalt, vanadium, and a few others.

[b]Monetary metal not included under uses.

[c]Antimony, cadmium, etc.

[d]According to U.S. Bureau of Mines, the World Bank, the National Commission on Materials, etc.

processed or only slightly processed. Australia, while processing a substantial portion of its mineral production, could obviously process more. The reason for this situation can be traced to such factors as the structure of ownership in the minerals industries, uneven technological development over most of the world, and the economics of distribution.

If we consider ownership, we see that in most of the mineral-producing industries, there are only a few large firms, and these firms prefer that the output of the mines they own, control, or have some special relationship with is processed in their smelters or refineries—even though these installations may be on opposite sides of the world from each other. In the case of a country like Australia, this practice has caused some agitation in that technically Australia possesses the know-how required to process a large part of its ore and at present needs the employment opportunities that would result from this process. Of course, the alteration in the world energy picture may be changing all of this. The processing of aluminum, for example, is a very energy-intensive operation, and because of the increase in energy prices, smelters in Japan, Western Europe, and even the United States have experienced considerable economic difficulties. This situation has resulted in the management of these smelters (who are still the managers and marketers for the semifabrication stage of operations in many integrated firms) becoming more friendly to the idea of locating their smelters in the energy-rich parts of the world.

One more important point needs to be taken up within the context of the present discussion. In table 8-1 we see a great increase in the value of production as we move from the mining to the metal-producing stage. Among other things, this increase has led many economists in ore-producing countries to insist that more metal-producing capacity is required in these countries. What needs to be appreciated here is that this high value—which represents income accruing to the employees of processing installations, profits going to the owners of these installations as well as undistributed profits, and tax revenues (with all of these coming under the heading of *value added*)—does not necessarily represent a larger *profit rate*. Thus private investors (firms or individuals) often claim that activities like smelting and refining possess too low a profit rate to be attractive, regardless of the *social profitability* associated with higher employment and more tax revenues. (Among these private investors can be mentioned some Australian mining firms who have argued, and perhaps correctly from their point of view, that any private money being invested in the mining industry should go to ore extraction rather than processing.) It is for this reason that in New South Wales the state government has become involved in attempts to attract processing facilities to the state. What this means is that they are willing to establish conditions that will make smelting or refining an interesting (that is, profitable) alternative for both Australian and foreign firms.

The final topic in this section will be introduced by table 8-2, which shows where the majority of the world's metals (by value) were being produced in

Table 8-2
Total Value of Nonfuel Mineral Production (1973)
(*In Millions of Dollars*)

	Total Value of Production (Millions of Dollars)	Average Yearly Growth of Production (%) 1950-1973	Most Important Mineral
1. Soviet Union	6800	7.0	–
2. United States	5100	4.0	–
3. Canada	3800	3.9	Copper, nickel, gold
4. South Africa	3800	6.5	Gold
5. Australia[a]	1700	8.8	Iron, bauxite, copper, nickel
6. Chile	1200	–	Copper
7. China	1200	–	Iron, copper, tin
8. Zambia	1000	–	Copper
9. Zaire	870	–	Copper
10. Peru	690	–	Copper
11. Brazil	640	–	Iron
12. Mexico	610	–	Copper, zinc, iron
13. France	490	–	–
14. India	420	–	Iron
15. The Phillippines	410	–	–
16. West Germany	400	–	–
17. Sweden	390	–	Iron
18. Poland	390	–	–
19. Japan	380	5.7	–
20. Yugoslavia	360	–	–
Others	6050	–	–

Source: USBM and World Bank documents. Various issues of the OECD biannual economic report. Also OECD country studies. F.E. Banks, *Scarcity, Energy, and Economic Progress* (Lexington, Mass.: Lexington Books, D.C. Heath, 1977).
[a]Including Papua New Guinea.

1974. As this table makes clear, five countries have a special position in the scheme of things where the production of minerals is concerned. It is also the case that the United States is still apparently the largest producer of minerals (to include fuels), but in nonfuel minerals it now occupies second place to the U.S.S.R. Moreover, Russia is probably close to self-sufficiency in minerals while the United States is increasing its reliance on imports.

Canada is the world's leading mineral producer on a per-capita basis, but the increase in mineral output over the last decade or two in Australia, and particularly in the mining of iron ore and bauxite, has been phenomenal by any standard. Countries that are scheduled to grow in importance in the near future are China and Brazil, and perhaps a few others; while places like Indonesia, Namibia, and even Saudi Arabia could figure prominently in the world mineral economy before the end of the century.

Secondary Materials, Substitution, and Scarcity

The materials that move directly from the mining to the smelting/refining stage (without a detour into the world of the consumer) are called *primary materials*. But there is another source of industrial raw materials, which is *recycling* the *new* scrap that is generated in processing or fabricating in the form of shavings and so on; or obtaining and recycling the *old* scrap that can be reclaimed from finished goods containing the particular material. Scrap in general is called *secondary* metal in the case of metals, or *secondary materials* when we mean any commodity (for example, paper).

In the United States about 140 million tons of solid waste is discarded every year, to include 3 billion pounds of aluminum (which is equal to the entire consumption of aluminum in the United States in 1958). When it is remembered that recycling aluminum requires only about 5% of the energy input per unit of new aluminum, it means that recovering this metal would save 20 billion kWh of electricity, which is 1% of the energy used in the United States. Unfortunately, at present the recyling of old and new scrap in the United States yields less than 20% of the yearly consumption of aluminum, although in the short run they could obtain much more of this metal from some form of recycling.

The saving of energy, by itself, makes recycling valuable, although the aesthetic effect of removing the litter and landfills that are eyesores in countries like Australia and the United States would represent a clear gain to almost everyone. More important, it may be possible to build an entire industry around recycling, creating thousands of self-sustaining jobs. At present the "recycling movement" and the technology for making it viable seems to be progressing about as rapidly in the United States as anywhere else in the world with firms like Raytheon, Gruman, Boeing, and American Can constructing plants for mining aluminum and other recyclables from refuse; and the world's largest aluminum company, ALCOA, providing a major market for secondary collection.

In the United States in 1977, 6 billion cans containing aluminum were turned in, and for an average of 17¢ per pound (where a pound is about 23 cans), collectors received $45 million. Moreover, at Ames, Iowa a complete resource-recovery facility is in operation. This is a plant costing $5.6 million whose daily intake is 150 tons of garbage. This refuse is shredded in the plant, and magnets sort out iron and steel. Aluminum is also removed by magnets and eventually falls in a bin. One of the products of this plant is an assortment of unrecyclable materials such as food, paper, scraps, plastics, and fabrics. These are ground into fine particles and piped as a fuel to an electricity-generating plant. There, every 2 lb of waste gives off the heat of 1 lb of coal.

Recycling is extremely important from a political as well as an economic point of view. One of the reasons for the effectiveness of OPEC is that in the short run there is no substitute for petroleum; nor is there a technique known or likely to be developed in the foreseeable future for the economical recycling of this product. On the other hand, there is so much scrap material available in most industrial countries that for a limited period, if it were necessary and if the recycling capacity existed, secondary materials could become the chief industrial feedstock for a number of the major industrial countries.

This leads us to the matter of substitution. Aluminum and iron are among the most plentiful components of the earth's crust, and together with plastics and biological materials (for example, wood), they are capable of replacing most of the other metals in the production of industrial and consumer goods—although it should be made clear that this substitution will be an expensive operation if it has to take place in the near future. Substitution is also possible in consumption since consumers are for the most part interested in the services they can obtain from their purchases, relative to price, rather than the composition of these goods; and therefore if the price of goods containing scarce materials increases, these consumers should normally divert their attention to products containing less expensive materials.

The "Real Price" and Scarcity

The "real price" of an industrial raw material is the amount of "something" that must be given up for this material. This "something" might be labor hours or perhaps other goods and services such as foodstuffs, machinery, and haircuts. For example, from 1960 to about 1969, the price of most grades of oil was essentially constant at about $1.88 per barrel. At the same time the price of almost all other types of goods was rising. Thus if we can imagine someone receiving his or her pay in oil, at a rate of one barrel of oil each year over that period, then each passing year would have seen a decrease in the purchasing power of that oil in the sense that the oil would have purchased less each year of these other goods. (Conversely, anyone wanting to buy oil could have done so with a smaller amount of other goods each year.) In other words, the real price of oil sank over this period.

Actually, the real price of oil has been decreasing over most of this century, even though its money price has been increasing; and the same is true for the money and real prices of most nonfuel minerals. As we know, however, the oil price increases of 1973–1974 broke this trend for oil, and although at the present time the real price of oil may be falling again, its level is still much higher than it was in 1972.

On the other hand, the real price of most nonfuel minerals has continued to fall, although questions are constantly being raised these days as to how long this can continue. The conventional argument is that if a mineral were to show signs of becoming scarce, its real price would have to rise. The reason is that as world demand expands, and ore grades thin, capital, labor, and energy costs will make it increasingly more expensive to obtain additional increments of output, and in the long run technical progress will find it impossible to alleviate this situation. Consequently, with the real cost rising, the real price will have to be adjusted up. Moreover, when some of the owners of this mineral come to understand this situation, they will reason that their own long-range profit-maximizing interests can be best served by leaving the resource in the ground instead of extracting it.

The logic here is simple. Suppose we have a mineral that does not cost anything to extract and whose price is appreciating at a rate of 15% a year. At the same time assume that the rate of interest is 10%. In this situation if the mineral is extracted the following year, it can be sold for 15% more than this year; while if it is extracted this year, sold, and the money from the sale put in a bank, this money would only appreciate by 10% over the year. Thus the incentive would be for the owner of the mineral to leave it in the ground, which would then reflect negatively on present supplies of the mineral, which in turn could cause a still faster appreciation in the price of the mineral. For instance, some of the directors of OPEC have said that in the mid 1980s it will be unnecessary to negotiate among themselves the price of oil. Their argument is that the growing physical scarcity of oil reserves in some oil-producing countries at that time will leave these countries with no choice but to lower their production, and thus the decrease in supplies to the market will cause the money price of oil to rise very rapidly. In addition, buyers in the industrial countries will rush to pay this price. The basic truth behind this type of reasoning can be seen at the present time (May 1979): With exports from Iran temporarily halted, the spot price of oil has climbed to almost double the official price, and there is no shortage of purchasers.

As mentioned earlier, up to the first part of the 1970s, and probably even now, there has been a decreasing real price for all the important nonfuel minerals, but there is a finite possibility that this trend could be broken for several minerals, beginning with copper, tin, and silver. Taking copper to begin with, a number of investigators claim to have detected a slowdown in the rate of the decline of its real price, beginning a number of years ago. To understand this slowdown, we should appreciate that one of the principal causes of a falling real cost is technical progress, which compensates for the decrease in ore grade by holding down the cost of the inputs needed to produce copper, such as capital, thus making *all* inputs more efficient in the performance of various operations. The average grade of copper ore being mined across the world has fallen from 13% (which is the content, by percent, of the copper in a unit of copper ore) to about 0.8% in the last 75-100 years; but there are few people who would claim that the real cost of copper has increased.

On the other hand, if we confine our attention to the United States, we see that the average ore grade now being mined in between 0.6% and 0.7%; and according to the present projections will be between 0.2% and 0.3% at the end of the century. This decline in content will entail having to handle three times as much waste material (rock and tailings) and will require an energy input per unit of output that is almost three times that being used today. It would be very surprising if technical progress could function in such a way over the coming 20 years that it could substantially decrease these figures or, for that matter, reduce energy or materials-handling costs by a great deal. What we should expect instead is a sharp rise in the per-unit cost of obtaining copper at many installations, which will be paralleled by an increase in its money price and, in all likelihood, an increase in its real price. We cannot say just when this will happen, but one thing is certain: It cannot be postponed indefinitely.

With tin, the problem seems to be a steady decrease in the amount of reserves relative to consumption (or what is known as the "reserve/consumption ratio") as well as the slowest increase in the amount of reserves discovered of all the major industrial raw materials. To understand the significance of this problem, it should be pointed out that just as it is an economic blunder to search for or develop reserves too far in advance of when they will be required, a small ratio of reserves to consumption is unsatisfactory in that it raises the possibility that a large part of the useful life of the mining and processing equipment might be wasted if the stock of reserves is used up too soon. In theory, the outcome of a situation like this should be a rise in the price of the mineral in order to reduce the demand for it, thereby ensuring that this equipment will be more economically utilized.

Somewhat less sophisticated, the low reserve/consumption ratio is a palpable measure of scarcity; and if it gets too low, it will indicate to some producers that it is in their interest to leave increasing amounts of this mineral in the ground to take advantage of the price increases that will be invoked by its scarcity. As for silver, there is also the matter of a fairly low reserve/consumption ratio and the interesting fact that silver occurs to a considerable extent as a by-product of lead and zinc. The growth rate of consumption of these latter two minerals is expected to decrease as a result of the slowdown in international economic activity that is predicted for the next decade which will almost certainly reflect on the availability of silver.

Other minerals that may be approaching a high degree of scarcity, to include increasing real prices, include gold, wolfram, and antimony. No attempt will be made here to discuss the situation of these items, but it might be useful to note the position of the U.S. Bureau of Mines on these and some other metals. According to the USBM, there could be a shortage of supplies of both silver and wolfram in the short run, while the real price of copper and gold will probably begin the climb right after the turn of the century. Some doubt is attached to the status of tin, lead, and zinc; but in general, the conclusion seems to be that if the reserves of these minerals that have not yet been discovered but that are expected to be discovered (that is, hypothetical and speculative reserves) turn out not to exist, then the real price of these materials will also increase early in the next century.

The Price of Nonfuel Minerals

In examining the price of most industrial raw materials, there are two prices that are relevant: the *long-run* and the *short-run* price. The long-run price is a trend price and is formed by the interaction of long-run supply and demand. For example, demand for aluminum (metal) is expanding at a rate of several percent a year while there has been a marked slowdown in the construction of aluminum smelters. Thus regardless of short-run fluctuations in the price of aluminum, the long-run picture at present indicates an increase in price as demand outruns supply. (Although, admittedly, this picture could be altered if a recession in the major industrial countries reduced the demand for aluminum.) Interestingly enough, this may mean a decrease in the price of bauxite, whose supply is expanding rather rapidly as new mines are opened in places like Brazil since demand for bauxite is a function of the amount of aluminum-smelting capacity in the world.

On the other hand, short-run prices can often be explained by speculation, the misinterpretation of washroom rumors, wishful thinking, and so on. To get some idea of how these prices move, the reader can consider the movement of nonfuel mineral prices through the world economic upheaval caused by the oil price increases of 1973–1974.

As the war in Vietnam came to its end, the slow but steady rise in the price of some of the more important industrial inputs began to level off, and some prices even began to decrease. However, only a short time later the international business cycle gained a fresh impetus, and since the rise in economic activity was, for one of the few occasions in modern times, synchronized across almost all the major noncommunist industrial countries, commodity prices headed up once more.

Although it is not often recognized, there was a monetary factor also influencing the price of many primary commodities. Thanks to the war in Vietnam and the budget deficits that had been used to finance this war, the American dollar has begun to lose its crediability. The result was a sustained deterioration of the value of the dollar vis-à-vis other currencies (which the reader will find described in some detail in my book *The International Economy: A Modern Approach*), as well as a growing speculation against the dollar in favor of various commodities (to include agricultural raw materials as well as minerals and metals).

Much of this confusion was in the process of being dissipated, largely by some major currency alignments, when the October War broke out between Egypt, Syria, and Israel. The war featured an oil embargo by the oil-producing Arab countries against several oil-importing countries deemed too friendly to Israel, and a 300% increase in price to the rest of their clientele. This embargo and the oil price increases raised the specter of a general shortage of raw materials; thus instead of a drastic fall in the demand for these materials, which could have been expected as a result of the sag in industrial production across the world, there was a huge increase in the demand for metals and minerals by inventory holders. This demand, which could be characterized as being of a precautionary and speculative nature, drove the price of commodities like copper and aluminum to record heights.

About the middle of 1974, with the inventories of most raw materials extremely high in relation to the demand for them as inputs in the current production process, businessmen, speculators, and even some economists began coming to their senses. The result was a crash in the price of industrial raw materials as inventory holders tried to get rid of their stocks in a world where there was no place for them to go except at bargain-basement prices. The onset of the 1974–1975 recession simply added fuel to the price decline, and since that time these prices have followed an irregular course. On several occasions since 1975 there have been mini buying panics, but in the case of, for example, copper, the price has stayed below the cost of production for an embarrassingly long time. Of course, even so, LDCs have generally been unwilling to decrease their production because their receipts from the sale of copper (and other minerals) constitute for the most part their access to hard currency, and this in turn puts a further downward pressure on prices.

Now, as the reader has undoubtedly noticed, this matter of stocks (that is, inventories) has come up again. They have been referred to earlier, but it might be interesting to see how they fit into the picture in regard to something like the mini buying panics referred to earlier. For example, suppose there is a nice balance between current demand and current supply of an industrial mineral or metal and as a result a fairly steady price. In the background there are, of course, stocks since both producers and consumers must protect themselves against interruptions in production or distribution, unexpected surges in demand, and so on; and in addition, there are some stocks being held by speculators who believe that the demand for the materials in their possession will increase, and as a result they will reap a substantial profit.

Let us also assume that a wave of optimism wafts across various parts of the economy, leading to a general belief that there will be an increase in demand for all types of consumer and investment goods and by extension the industrial raw materials that go into their production. In addition, these bullish beliefs are embraced by a number of individuals and organizations disposing over substantial amounts of purchasing power. What we are likely to get now is an increase in the price of the industrial metals; however, we should note very carefully just how this price increase takes place.

Initially, with demand and supply balanced, anyone buying raw materials will have to buy them from someone's inventories. This *can* happen because there are some inventory holders who feel more strongly about the possibility of an economic upturn than other inventory holders, and they make these feelings known in the usual fashion. This type of buying, in addition to some frenzied ordering in which all sorts of premiums are offered producers if they will accelerate deliveries, serves to raise the general price level of raw materials and eventually causes an increase in *flow* supply (that is, current production). Remember though that our assumption was that this business-cycle upswing was predicted for the future, which is when these raw materials will be needed in the production process; thus to begin with, they are purchased as an addition to inventories.

The issue now turns on what actually happens with the economic situation. *If* there is a general increase in economic activity, then these inventories will be needed, and their owners will be judged wise and farsighted. However, if, as happened following the "buying panics" of 1976 and 1978, no major economic upswing materializes, then these additional inventories—relative to the current consumption of raw materials in industrial and consumption activities—will be judged excessive, and their owners will try to get rid of them. This action would normally result in a downward pressure on the market price of the raw material. It is, as a matter of fact, waves of bullishness and bearishness, often initiated by what amounts to no more than idle fancy, rather than a detailed analysis of the prevailing economic situation, that explain the ripples that we see on the price charts in the business press and that, taken by themselves, need not have any logical relation to the realities of supply and demand.

Some comment is now necessary on one of the more pedestrian aspects of the nonfuel minerals markets. The pricing of a number of metals follows a scheme known as *producer* or *posted* pricing. What this involves is the producers' of these metals—or, to be specific, the largest producers of these metals—setting the price of these metals, generally with the approval of the authorities. Ideally, this price would cover their costs and provide them with a reasonable profit, where "reasonable" means not too low but at the same time not so high as to tempt other producers into the same line of activity.

It is also possible to identify another price that is called a *free market, open market, dealer,* or *merchant price.* If, for instance, during a certain period demand is appreciably higher than production, then in the short run this demand will be satisfied mostly by a reduction stocks. Some of these stocks will be held by producers or consumers of the commodity, but a considerable amount will be in the possession of individuals (or firms) called *merchants* or *dealers.* As a rule, merchants buy and sell commodities within the main producer-consumer channels, and only on rare occasions will they acquire production facilities. They *do* participate in arbitrage operations, buy scrap materials for refineries processing secondary materials, and act as agents for both buyers and sellers on the various commodity exchanges, and they are particularly active in the financing, holding, and selling of stocks. In line with the previous discussion, these stocks would not be sold at the producer price but at a so-called free market or dealer price that, unlike the more inflexible producer price, is highly responsive to supply and demand. (As the reader can see in figure 5-4, the producer price has a tendency to move in "steps.")

On more aspect of this topic should be considered. The producer price and the free market price sometimes differ by considerable amounts, and it has been noticed that in some cases in which the free market price is much higher than the producer price, it can take a great deal of time before producers raise the producer price to this market price. The explanation here turns on the concept of long-run as opposed to short-run profit maximization. A high producer price over an extended period might provoke minor producers to expand their capacity and perhaps bring some new producers into the field. As a result the practice has been for the largest producers of the major raw materials to seek a "target profit" via a pricing policy that involves stable prices for rather long periods rather than the constant ups and downs of the free market. Some of this may be changing, however, since the sight of the free market price staying at a much higher level than the producer price has caused a few large producers (most notably in copper) to break ranks with their colleagues and price their products at the free market price. (Much of this discussion applies to the world oil market, where the "producer" price is the OPEC price, and the free market is the Rotterdam spot market.)

The London Metal Exchange

In speaking of a free market price, the question immediately arises as to where this price appears. This brings us immediately to the metal exchanges, in particular the New York Commodity Exchange (COMEX) and the London Metal Exchange (LME). Of the two, the LME is generally considered the most important in terms of turnover, physical deliveries, and influence on the price of metals in general—which is why it will be looked at here; but COMEX handles a wider assortment of metals and, unlike the LME, also provides facilities for trading hides and rubber.

The LME occupies the site of the old Roman Forum in London. Each day there are four market meetings, two in the "morning" (beginning at noon) and two in the afternoon, beginning at 3:40. These meetings are broken down into sessions of five minutes, during which time 30 LME "ring-member" dealers buy and sell copper, silver, tin, lead, and zinc. (Nickel and aluminum will soon be included).

They buy and sell for their clients the world over who are adjusting their inventories in one direction or another, getting rid of surplus metal or acquiring more metal immediately to service unexpected orders. (Note that dealings on the LME involve metals, not ores, for which precise specifications can be established.) The price established in these sessions therefore indicates the state of world wide supply and demand to a remarkable extent, which is why many contracts in metal trading are based on these prices. For example, copper producers in many countries sell their copper for forward delivery at a price related to the price prevailing on the LME on or around the date at which they are to deliver this copper to their client. This might be called *exchange pricing,* and here we have the interesting, and by textbooks unanalyzed situation, of selling a known quantity at an unknown price.

An equally important function of the LME is to provide facilities for hedging, or transferring price risk from the buyers and sellers of physical commodities to speculators. There is a detailed, nontechnical discussion of this process in chapter 6 of this book, complete with examples, and the reader is referred to that chapter; but it can be mentioned here that the medium for transferring these price risks is provided by *futures contracts,* and the guiding rule is as follows: *Those wishing to insure against a decrease in price sell futures; those wishing to insure against an increase in price buy futures.*

Whom do they buy from? They buy from speculators who are betting that the price of a given commodity rises or falls. What is a futures contract? Strictly speaking, a futures contract is a forward contract in that it specifies physical delivery; but delivery can be avoided because the contract can be offset. "Offsetting" it can take place as follows: a producer sells a physical commodity for forward delivery at a price related to the price of the commodity on a metal exchange at the time of delivery. At the same time he or she *sells* a futures contract. Then at the time of delivery the producer *buys* a futures contract, offsetting the previous sale. If the price of the physical commodity has fallen, the producer loses on the physical transaction; but if the price of the futures contract also falls—as it normally should—then the producer will gain on the futures or "paper" transaction. (Again the manner in which the price of futures contracts move in relation to the price of the commodity is explained in detail in chapter 6.)

It is also of interest to note, given the importance of inventories in the metals markets, that the inventories in the LME warehouses (to and from which all LME transactions resulting in physical deliveries must take place) seem to be directly related to the market price of various metals. Thus at the time when this book was being completed, there was a kind of *hausse* (rise) on the world copper market with inventories in LME warehouses falling rapidly. These stocks are not disppearing from the face of the earth, however, but are moving to the warehouses of speculators, consumers, and producers who think that because of the deterioration of the situation in southern Africa, as well as some strikes in South America, copper will soon be hard to come by. Figure 8–2 shows the relation between stocks in LME warehouses and the price of copper.

One final comment is in order here. As far as I know, to be elected to a seat on the LME, pedigree counts as much as financial backing. Up to now, at least, no graduates of Roosevelt University or the ranks of the 24th Infantry Regiment have penetrated the hallowed inner chambers of the exchange; the decor is almost exclusively that associated with the *right* school, university, and regiment. The theory here, apparently, is that when agreements involving huge sums of money are being made on a nod of the head or the shake of a hand, it is best that gentlemen are involved—although, regrettably, even gentlemen can suffer from

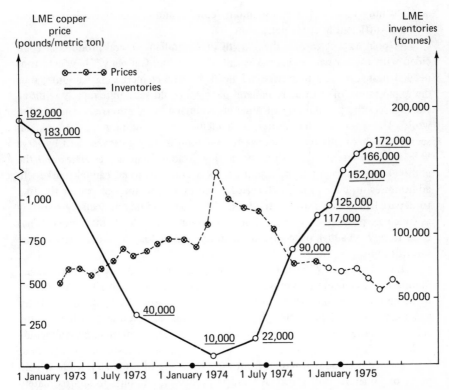

Figure 8-2. Prices and Inventories on the London Metal Exchange

lapses of memory when their bank accounts become anemic. Film buffs can get a glimpse of the session activities in the movie "The Silver Bears." Whether the actor portraying a ring dealer attended Eton or served in the Brigade of Guards is unknown to me, although everything considered, I have my doubts; but even if he was not born with a silver spoon in his mouth he certainly gives the impression of being able to put his heart and soul into the buying and selling of this commodity.

Minerals and Australia

Australia is today one of the most dynamic mineral producers in the world, if not the most dynamic. This country's rise to status as a mineral producer began in the early 1960s with the opening up of some new iron-ore deposits in the Pilbara and the beginning of bauxite mining and processing based on the ample deposits of Queensland, Western Australia, and the Northern Territory. Australia also has huge coal, nickel, and uranium deposits, and in addition, it is more than 65% self-sufficient in crude petroleum.

As pointed out earlier, the growth of Australian mineral production coincided with the postwar industrial ascent of Japan and Germany. These countries needed industrial raw materials and needed them in increasing amounts; thus the exploitation of Australian mineral deposits could accelerate rapidly without endangering the profitability of mineral-producing installations elsewhere in the world. There was also the matter of financing the expanding Australian mineral sector. Much of this took place via the medium of foreign investment by large U.S. and United Kingdom firms possessing sizable financial resources *and* the ability to complement these resources on the international capital market. In addition, as noted previously, the Australian mining sector was one of the first to depart from the worldwide tradition of financing mining projects by resorting as much as possible to internal funds and equities. Instead, the tendency has been to rely on a high proportion of debt secured by long-term contracts for the sale of output.

According to D.W. Barnett (1978), new capital expenditure in the Australian mining industry came to almost $6 billion between 1964 and 1976. About half of this money was imported into the country, but Australians still own slightly less than half of their mining industry and "control" even less—at least in the economic sense. As the situation now stands, the mining of minerals (to include fuels) contributes about 4.5% to the Australian gross domestic product (with at least the same amount resulting from the processing of minerals); while the exports of minerals, both processed and unprocessed, amounts to approximately 35% of the value of Australian exports. These figures can be compared to figures of 1.87 and 6.56, respectively, for 1960. Some other trade figures for Australia are given in table 8-3; this table also shows the Australian reserves of some of the most important minerals as well as some information about ownership.

The previous discussion provides a background to a brief examination of the Australian mining industry and, by extension, the Australian economy as a whole. Moreover, this discussion is relevant for many mineral-exporting countries, to include those in the less developed world. If we take the matter of ownership, the Australian capital market is not in its natural state large enough to ensure a gradual transfer of the Australian mining industry to Australian hands. Moreover, there is hardly any possibility of *widening* it through increasing the number of savers or *deepening* it through increasing per-capita savings. Probably the most interesting prospect would be the formation of one or more Australian development banks which, drawing on the almost unlimited resources that the Euromarket is capable of mobilizing at the present time, could finance a large part of the mining and mineral-processing sector that cannot be accommodated by the local capital market. In the eyes of the London, New York, Zurich, and Frankfort bankers who are scurrying from one corner of the globe to another to find borrowers for their growing stocks of eurocurrencies, Australia undoubtedly has a prime credit rating, and the $1.5-$2 billion per year that the Australian mining industry will probably require in the near future is small beer to these gentlemen.

The next topic has to do with the effect of an expansion of the mining sector on the rest of the economy. It has always been clear that the multiplier effect of the mining industry is low unless this industry results in a great deal of further processing. In other words, a situation that involved an expansion of mining (without processing) would, in favorable market circumstances, mean a nice profit for mine owners but not much in the way of development for the rest of the economy. As emphasized earlier, the change in the world energy situation is gradually swinging the comparative advantage in further processing to Australia, particularly in the production of aluminum metal; but even so it must be remembered that smelting and refining are mainly capital and energy intensive, rather than labor intensive, and thus the employment opportunities created by futher processing will be to a considerable extent indirect (via buying from other industries) rather than direct. Australians might therefore feel that it would be best for them for this "buying" to take place in Australia rather than abroad; and for this to happen, these "other industries" would either have to exist in Australia or be created in time to tie into an expanding processing sector. Then too it might be considered appropriate for the large amount of nonlabor income in the processing sector to stay in Australia. For this to happen, Australians would have to own the greater part of these processing facilities.

When the matter of establishing industrial installations in Australia is raised, certain other topics immediately become relevant. The principal topic here concerns the private and social profitability of a large Australian processing sector in a highly competitive world, given the protectionistic measures favored by Australian governments (and various groups within the Australian economy) and the well-known aggressiveness of Australian labor unions. Taking the last theme first, the attitude of the Australian unions is simply a reflection of the extraordinary degree of materialism that characterizes Australian society as a whole and, considering the extraordinary possibilities offered by what has been called the "lucky country," could hardly be otherwise. Besides, it seems pointless to criticize Australian labor unions for attempting to increase the earnings of their members when other groups in the community, already privileged by *any* standard of economic comparison, are able to ensure a continued improvement in their direct and indirect remuneration regardless of the state of the economy (where "indirect remuneration" refers to, for example, working conditions and options concerning working hours). Conditions in Australia are, in fact, very similar to those in Sweden in that certain groups in the community, both white and blue collar, are able to enjoy a favorable economic outlook at the expense of the unemployed, school dropouts, people on pensions and fixed incomes, and of course the socially and physically handicapped. In this kind of ambiance, it becomes natural for everyone, to include trade unionists, to make it their business to get as much as they can as soon as they can.

As for the matter of protectionism, Professor Ryokichi Hirono of Seiku

Table 8-3
Some Data on the Australian Mining Industry

	Exports[d] (Millions of Dollars)	Reserves[b]		Industrial Production[c]		Percent of Value Added Going to Foreigners
		Australia	World	Australia	World	
Coal (million tonnes)	1,290.5	32,100	765,000	97	3,110	63.7
Liquid petroleum gas (mil · m³)	151.3	815	72,010	4.8	1,345	61.1
Iron ore (million tonnes)	972.2	17,800	253,000	97	850	
Pig iron	51.1					
Crude iron and steel	201.4					
Nickel concentrates	13.6					
Nickel matte	179.5					35.6[a]
Nickel unwrought	47.0					
Nickel ore (000 tonnes)	52.3	2,000	45,000	46	730	
Manganese ore (million tonnes)	18.5	490	5,443	1.5	23	
Tungsten Concentrates						
Tungsten ore (000 tonnes)		77	2,060	2.2	47	
Bauxite (million tons)	62.4	3,000	15,500	20	80	68.0
Alumina	549.9					
Aluminum	55.5					
Copper ore (000 tonnes)	51.4	6,000	440,000	250	7,910	51.2[a]
Copper (unrefined)	17.0					

Copper refined	104.3					
Zinc ores (000 tonnes)	78.4	19,000	119,000	463	5,690	
Zinc and zinc alloys	89.3					
Lead ores (000 tonnes)	24.0	14,000	145,000	370	3,550	
Lead and lead alloys	139.7					
Tin concentrates	17.9					34.6[a]
Tin unwrought	15.4					
Tin ore (000 tonnes)	82.7	332	10,160	10	180	
Rutile (000 tonnes)		9,200	13,000	319	350	
Ilmenite (000 tonnes)	10.6	58,400	570,000	817	3,860	
Zircon (000 tonnes)	62.5	15,700	33,000	368	530	
Industrial salt (million tonnes)	28.5		–	4.9	159	
Crude oil (m. cubic meters)		387	116,064	23	3,220	61
Uranium (000 tonnes)		312	2,022	–	19	23
Total (minerals + metals)	4,366 U.S. dollars					
Total (all australian exports)	12,103 U.S. dollars					

Source: Australian Bureau of Mineral Resources; U.S. Bureau of Mines; U.N. publications; and R.B. McKern, *Multinational Enterprises and Natural Resources* (Sydney: McGraw-Hill, 1976).

[a] Applies to the entire industry.
[b] For the year 1975.
[c] For the year 1974.
[d] Average of the period 1975–1976.

University (Tokyo), on a recent trip to Australia, criticized the Australian habit of protecting high wages and jobs with tariffs that can be passed on to the consumer. Resolving this issue probably reduces to a matter of taste, by which the taste of this humble economist leans toward the Japanese system—a system that does *not,* as Professor Hirono has implied, encourage the Adam Smith type of free competition, but rather the government's entering into a kind of partnership with industry and supporting those firms and sectors that it feels have a future. Of course, for a strategy of this type to work, the people of a country must endeavor to ensure that both their politicans and senior civil servants are neither elected nor promoted beyond their intellectual and moral capacities, which is something that the electorate in all except a few countries seem to have lost interest in doing.

Another type of partnership that might be relevant here involves unions and management at the corporate level, which, as in the case of West Germany, could include some degree of *mitbestimmung;* and thus wage bargaining would take into consideration the market situation for the particular product. The last suggestion, however, is strictly of the ivory-tower variety since a comprehensive discussion of the market situation would involve taking up productivity—which is a subject that is strictly taboo in Australia, just as it is in almost all industrial countries.

Next we come to an interesting analysis of the Australian mining sector by Robert Gregory (1976). Reduced to its simplest possible terms and ignoring some of its extensions and embellishments, Gregory's model envisages a situation where a large-scale expansion of the mining sector would draw resources into the mining sector and away from the rest of the economy. Since minables are exported, there would also be an upward pressure on the Australian exchange rate that would work to the disadvantage of other exporters in that if the Australian exchange rate appreciated or were revalued, the price of Australian goods would increase to foreigners. A problem would also be created for other exporters and producers of nontradables by the increase in the cost of various production factors due to a greater demand for them by the mining sector.

There is no point in looking at the analytics of Gregory's model in this nontechnical survey, although personally I am prepared to argue that the techniques he employs are greatly superior to those his critics use (and which involve simple two sector models without the slightest relevance to the real world). But even so, on the basis of my own limited knowledge of the Australian economy, I believe that economic upswings, or booms, have tended to strengthen the economy in both the short *and* long run, regardless of what they are based on. What they do is bring in foreign capital as well as immigrants willing to extend themselves both physically and mentally to gain a nice helping of prosperity. In addition, the morale of both employers and employees is raised, and last but not least, many people come into possession of bank accounts, durables, and property that will be useful in less gratifying times.

At the same time attention should be given to the fact that the phenomenon Gregory has attempted to capture in his model has to a certain extent appeared in Norway. It was and is my belief that Norwegian oil can pay its own way (and the same is true of Australian minerals), but this does not happen automatically. What the expanding oil economy did in Norway was place an intolerable pressure on the rest of the export sector, though not quite in the way Gregory envisages in his model. The principal villain in Norway was cost inflation, which led to price inflation, specifically the inflation of export prices. The Norwegian government, quite naturally, did not understand the exact nature of the impact that oil would have on their economy, and even if they had understood, their intellectual reflexes were not limber enough for them to do what they should have done about it. What they ended up by doing, naturally, was devaluing the Norwegian krona.

But even so, there is no point in claiming that Norway has been made worse off because of its oil boom. On the contrary, even with inflation, psychological traumas of one type or another, and the whine of indolent scribblers concerned with the passing of traditional values (which actually began to pass slightly more than a decade ago), Norway has become an island of security in a Scandinavia where incompetent governments, obsessed with trivia, preside over what may be the twilight of a golden age of social and economic progress.

Finally, it must be underlined one again that much of what has been considered previously merely serves to obscure an even more important controversy: Many Australians have come to think of their mineral wealth as being almost infinite. It may be, but not on the basis of the most optimistic estimates of reserves available today. Thus sooner or later someone must answer some questions dealing with just what part minerals should play in the economic life of Australia in a few decades, instead of just a few years, and preferably this question should be asked before accelerated depletion leaves only one possible answer. None.

Bibliography and Note on the Literature

All readers who take their natural resource economics seriously should make an effort to read the quarterly *Resources Policy,* published by the IPC Science and Technology Press in London. This is by far the most important periodical in the resource field. In addition, the IPC press publishes the most important of the energy journals, *Energy Policy.* First-class reviews with an engineering bias are the *Engineering and Mining Journal,* and the *Canadian Mining and Metallurgical Bulletin.* Of the academic journals, I have found the *Journal of World Trade Law* the most useful for the topics treated in this book. Another important publication is the bimonthly *Intereconomics.*

For an interesting and useful background to natural resource and energy economics, the reader should consult the book edited by David Pearce (1976). There is also important material on recycling in Pearce's book *Environmental Economics.* My own books *The Economics of Natural Resources* and *Scarcity, Energy, and Economic Progress* also treat energy and natural resources in some detail. There is an excellent survey of these matters by Anthony Fisher and F.M. Peterson (1974). I also think that it would be illuminating for the reader to examine the books of Wolfgang Gluschke, Joseph Shaw, and Bension Varon (1979), P. Connelly and Robert Perlman (1975), Raymond Mikesell (1975), and R.B. McKern (1976).

Some of the best work on aluminum has been done at the World Bank. Charles River Associates, a Boston consulting firm; the London-based Commodity Research Unit (CRU); and the Paris-based Eurofinance (or Euroeconomics) have also made important investigations of the aluminum market. In addition, CRU has made some impressive investigations of the world copper market. For an introduction to the theory of exhaustible resources, the basic approach is outlined in the papers of Solow (1974, 1975) and Nordhaus (1973, 1974), but the most important extensions of this work are now being carried out by Murray Kemp and Michael Hoel.

Where market theory applicable to this field is concerned, the reader should consult the work of Robert Clower and Robert Pindyck. The latter is also rather fluent on cartel policy. Other useful references here are Van Duyne (1975), Amacher and Sweeney (1976), David McNicol (1978), and my book *The International Economy: A Modern Approach.* The work of Professor Ingo Walter should be read for information about underwater resources. The reader who is particularly interested in natural resource and energy problems in Australia can consult the work of Michael Folie, especially Folie and Gregory McColl (1978); the activities of the Centre for Resource and Environmental Studies of the Australian National University, headed by Professor Stuart Harris; and the

Centre for Economic Research of the University of New South Wales under John Nevile, who is professor at that institution.

Energy is a very important subject where nonfuel minerals are concerned, and here the most important references for an overall theoretical examination are Slesser (1978) and Earl Cook (1976). In the matter of econometrics I would like to cite the work of W. Witherell, M. Desai, Hans J. Timm, and W. Labys; but the most imaginative econometric model of a commodity market is still that of Franklin Fisher and Associates (1972). It should be emphasized, however, that a great deal of commodity econometrics is basically useless; and purchasers of this type of econometric output should have it clear for themselves that econometrics is *not* a science, and they are paying for a product that no sober scientist could possibly take seriously in the form in which it is often sold.

Going to the subject of the taxation of resources, I can recommend a survey by Richard Dowell. In particular he looks at an interesting new approach to this matter by Ross Garnault. A broad range of topics in resource economics have been well handled by Paul Bradley, Lawrence Copithorne, and the Govetts. (He is a geologist and she is a minerals economist).

A large part of the commodity debate has been neatly summed up by Hugh Corbet (1975). Kenji Takeuchi, Jos de Vries, Marian Radetzki, Juergen Donges, and many others are making noteworthy contributions on the topic of cartels, commodity agreements, and commodity politics. The same is true of Rachel McCulloch; and the late Professor Harry G. Johnston had a great many provocative things to say about the systematic misuse of economic theory by certain international organizations. Finally, Jay Colebrooke of UNCTAD has been responsible for presenting a great deal of valuable material on the stockpiling of primary commodities; and for the reader looking for a good background text in economic theory that is applicable to the topics taken up in this book, I suggest that he or she should inspect the intermediate textbook of James Quirk (1976).

Alchian, A., and Allen, W. *Exchange and Production Theory in Use.* Belmont, Calif.: Wadsworth, 1964.

Amacher, Ryan C., and Sweeney, Richard J. "International Commodity Cartels and the Threat of New Entry: Implications of Ocean Mineral Resources." *Kyklos* 29 (1976).

Arrow, Kenneth. *Information and Economic Theory.* Stockholm: Federation of Swedish Industries, 1975.

Bailly, P.A. "The Problems of Converting Resources to Reserves." *Mining Engineering* (1976).

Bambrick, Susan. *Australian Minerals and Energy Policy.* Canberra: Australian National University Press, 1979.

Banks, Ferdinand E. "An Econometric Model of the World Tin Economy: A Comment. *Econometrica,* July 1972.

——. "A Note on Some Theoretical Issues of Resource Depletion." *The Journal of Economic Theory,* October 1974*a.*

——. *The World Copper Market: An Economic Analysis.* Boston: Ballinger, 1974*b.*

——. "Elementary Investment Theory: An Optimal Control Approach." *Jahrbucher für Nationalekonomi und Statistik* (1975).

——. "The Economics and Politics of Primary Commodities." *Journal of World Trade Law,* November/December 1976*a.*

——. *The Economics of Natural Resources.* New York: Plenum, 1976*b.*

——. "Laying Hamlets Ghost in Commodities." *New Scientist,* August 1976*c.*

——. *Scarcity, Energy, and Economic Progress.* Lexington, Mass.: Lexington Books, D.C. Heath, 1977.

——. *The International Economy: A Modern Approach.* Lexington, Mass: Lexington Books, D.C. Heath, 1979*a.*

——. "The 'New' Economics of Iron and Steel." *Resources Policy* (1979*b.*

Barnet, H.J., and Morse, C. *Scarcity and Growth.* Baltimore: Johns Hopkins University Press, 1963.

Barnett, D.W. *Minerals and Energy: The Impact upon the Australian Environment.* Unpublished manuscript. Sydney: Macquarie University, 1978.

Beckerman, Wilfred. "Economists, Scientists, and Environmental Catastrophe." *Oxford Economic Papers,* November 1972.

Berge, Helge. "En Ny Ekonomisk Världsordning?" *Ekonomisk Revy,* March 1977.

Bergsten, C. Fred. "A New OPEC in Bauxite." *Challenge,* July/August 1976.

Billerbeck, K. "On Negotiating a New World Order of the World Copper Market." *Occasional Paper of the German Development Institute,* no. 33, 1975.

Bosson, Rex, and Varon, Bension. *The Mining Industry and the Developing Countries.* New York: Oxford University Press, 1977.

Bradley, Paul G. *The Economics of Crude Petroleum Production.* Amsterdam: North Holland, 1967.

Bradshaw, Thornton. "My Case for National Planning." *Fortune,* February 1977.

Brooks, David. *Supply and Competition in Minor Metals.* Baltimore: Johns Hopkins University Press, 1965.

Brooks, David, and Andrews, P.W. "Mineral Resources, Economic Growth, and World Population." *Science,* July 5, 1974.

Brown, Lester. "Rich Countries and Poor in a Finite Interdependent World." In *The No Growth Society,* edited by Mancur Olson and Hans H. Landsberg. New York: W.H. Norton, 1973.

Brown, M., and Butler, J. *The Production, Marketing, and Consumption of Copper and Aluminum.* New York: Praeger, 1968.

Brubaker, S. *Trends in the World Aluminum Industry.* Baltimore: Johns Hopkins University Press, 1967.

Bushaw, D.W., and Clower, R.W. *Introduction to Mathematical Economics.* Homewood, Ill.: Irwin, 1957.

Carman, J. "Notes and Observations on Foreign Mineral Ventures." *Mining Engineering,* September, 1967.

Charles River Associates. *Policy Implications on Producer Country Supply Restrictions: The World Aluminum/Bauxite Market.* Cambridge, Mass., 1977.

Chapman, Peter. *Fuel's Paradise.* Harmondsworth: Penguin Books, 1975.

Clower, Robert. "An Investigation into the Dynamics of Investment." *American Economic Review* (1954).

Clower, Robert, and Due, J.F. *Microeconomics.* Homewood, Ill.: Irwin, 1972.

Connelly, Phillip, and Perlman, R. *The Politics of Scarcity.* London: Oxford University Press, 1975.

Cook, Earl. "Limits to the Exploitation of Nonrenewable Resources." *Science,* 20 February 1976.

———. *Man, Energy, Society.* San Francisco: W.H. Freeman and Co., 1976.

Cooper, Richard, and Lawrence, R.Z. "The 1972-75 Commodity Boom." *Brookings Papers on Economic Activity,* no. 3, 1975.

Corbet, Hugh. *Raw Materials: Beyond the Rhetoric of Commodity Power.* London: Trade Policy Research Centre, 1975.

Copithorne, L.W. "The Role of Gold and Base Metals in Regional Economic Activity." Economic Council of Canada, 1976.

Cranston, D.A., and Martin, H.C. "Are Ore Discovery Costs Increasing?" *Canadian Mining Journal* (1973).

Desai, M. "An Econometric Model of the World Tin Economy, 1948-61." *Econometrica* (1966).

———. "An Econometric Model of the World Tin Economy: A Reply to Mr. Banks." *Econometrica* (1972).

Donges, Juergen B. "UNCTAD's Integration Program for Commodities." *Resources Policy,* March 1979.

Dorr, A. "International Trade in the Primary Aluminum Industry." Ph.D. Thesis, Pennsylvania State University, 1975.

Dunham, Kingsley. "How Long Will Our Minerals Last?" *New Scientist,* January 1974.

Dowell, Richard. "Resources Rent Taxation." Stencil. The Australian Graduate School of Management, 1978.

Ertek, Tumay. "The World Demand for Copper, 1948-63: An Econometric Study." Ph.D. dissertation, University of Wisconsin, 1967.

Farin, P., and Reibsamen, G. "Aluminum: Profile of an Industry." *Metals Week* (1969).

Finger, J.M., and Kreinin, M. "A Critical Survey of the New International Economic Order." *Journal of World Trade Law,* November/December 1976.

Fischman, L.L., and Landsberg, H.H. "Adequacy of Non-fuel Minerals and

Forest Resources." In *Population, Resources, and the Environment,* edited by R.G. Ridker. Washington, D.C.: U.S. Government Printing Office, 1972.

Fisher, Anthony C. "On Measures of Natural Resources Scarcity." *International Institute for Applied Systems Analysis,* February 1977.

Fisher, Anthony C., and Petersen, Frederic M. "Natural Resources and the Environment in Economics." Mimeographed. University of Maryland, 1974.

Fisher, Franklin M.; Cootner, P.H.; and Baily, M.N.; "An Econometric Model of the World Copper Industry." *Bell Journal of Economics,* Autumn 1972.

Folie, Michael, and McColl, Gregory. The International Energy Situation Five Years After the OPEC Price Rises (Center for Economic Research, Sidney), 1971.

Forrester, Jay W. *World Dynamics.* Cambridge, Mass.: Wright-Allen, 1971.

Fox, William A. *The Working of A Tin Agreement.* London: Mining Journal Books, 1974.

Fuller, Carlos R. "The Future Development of World Copper Mining." *CIPEC Quarterly Review,* July/September 1976.

Garnault, Ross. "Australian Trade with Southeast Asia." Ph.D. dissertation, The Australian National University, 1972.

Garnault, Ross, and Clunies-Ross, A. "Uncertainty, Risk Aversion, and the Taxing of Natural Resource Projects." *Economic Journal,* June 1975.

Gluschke, Wolfgang, Shaw, Joseph and Varon, Bension. *Copper: The Next Fifteen Years* Dordrecht: D. Riedel Publishing Co. 1979.

Govett, M.H., and Covett, G.J.S. *World Mineral Supplies: Assessment and Perspective.* New York: Elsevier, 1976.

Gregory, Robert. "Some Implications of Growth in the Minerals Sector." *Australian Journal of Agricultural Economics* (1976).

Grillo, H. "The Importance of Scrap." *The Metal Bulletin,* Special Issue on Copper, 1965.

Habenicht, Horst. "Processing Mineral Raw Materials." *Intereconomica* 9/10 (1977).

Hargrease, David. "Feasibility Studies Outline: Cerro Colorado Development," *Mining Magazine,* August 1974.

Harlinger, Hildegard, "Neue Modelle für die Zukunft der Menschheit." IFO Institute für Wirtschaftforschung, February 1975.

Hashimoto, H. *Market Prospects for Aluminum and Bauxite.* The World Bank, 1978.

Herfindahl, O.C. *Copper Costs and Prices: 1879-1957.* Baltimore: Johns Hopkins University Press, 1959.

Herin, Jan, and Wijkman, Per M. *Den Internationella Bakgrunden.* Stockholm: Institut für Internationella Ekonomi, 1976.

Hotelling, Harold. "The Economics of Exhaustible Resources." *Journal of Political Economy,* April 1931.

Huang, A.C. *Prospects for World Import Demand for Bauxite, Alumina, and Aluminum from Developing Countries in the Seventies.* The Work Bank, 1974.

Hughes, Helen, and Singh, Shamsher. "Economic Rent: Incidence in Selected Metals and Minerals Resources." *Resources Policy,* June 1978.

International Economic Studies Institute. *Raw Materials and Foreign Policy.* Washington, D.C., 1976.

Jay, J.A., and Mirrlees, "The Desirability of Natural Resource Depletion." In *The Economics of Natural Resource Depletion,* edited by D.W. Pearce. London: Macmillan, 1975.

Johnson, Charles J. "Cartels in Minerals and Metals Supply." *Mining Congress Journal,* vol. 62, 1976.

Johnson, Harry. "Commodities: Less Developed Countries' Demand and Developed Countries' Response." Unpublished manuscript, 1976.

Jorgenson, Dale, and Hudson, Edward. "Economic Analysis of Alternative Energy Growth Patterns, 1975-2000." In *A Time to Choose,* edited by D. Freeman et al. Cambridge, Mass: Ballinger, 1974.

Kapitza, Peter. "Physics and the Energy Problem." *New Scientist,* October 1976.

Kellog, Herbert H. "Sizing Up the Energy Requirements for Producing Primary Metals." *Engineering and Mining Journal,* April 1977.

Köhler, Klaus. "Rohstoffpreisindizes: Methodik und Aussagefähigkeit." *Bremer Ausschuss für Wirtschaftforschung* 20 (1976).

Kolbe, H., and Timm, Hans J. Die Bestimmungsfaktorn der Preisentwicklung auf dem Weltmarkt fur Naturkautschuk—Eine Ökonometrische Modellanalyse." no. 10. Hamburg: HWWA Inst. für Wirtschaftsforschung, 1972.

Koyck, L.M. *Distributed Lags and Investment Analysis.* Amsterdam: North Holland, 1954.

Labys, Walter C., and Granger, C.W. *Speculation, Hedging, and Commodity Price Forecasts.* Lexington, Mass.: Lexington Books, D.C. Heath, 1970.

Labys, Walter C.; Rees, W.C.; and Elliott, C.M. "Copper Price Behavior and the London Metal Exchange." *Applied Economics* (1971).

LaQue, F.L. "Deep Ocean Mining: Prospects and Anticipated Short Term Benefits." The Center for the Study of Democratic Institutions, June 1970.

Laulajainen, Risto. "The U.S. Frash Industry, 1955-1975." *Sulphur,* March/April 1977.

Lave, Lester B., and Seskin, E.P. "Air Pollution and Human Health." *Science,* August 1970.

Leontief, W. *The Future of the World Economy.* New York: United Nations, 1977.

Lovering, T.S. "Mineral Resources from the Land." In *Resources and Man,* edited by Preston Cloud. San Francisco: W.H. Freeman, 1969.

Lowell, J.D. "Copper Resources in 1970." *Mining Engineering,* April 1970.

Malenbaum, Wilfred. *Material Requirements in the U.S. and Abroad in the Year 2000.* National Technical Information Service, Report PB 219-6/5/PB, 1977.

Malinvaud, E. *Lectures on Microeconomic Theory.* Amsterdam: North Holland, 1972.

Malmgren, Harold B. "The Raw Material and Commodity Controversy." Washington D.C.: International Economic Studies Institute, October 1975.

Manners, Gerald. *The Changing World Market for Iron Ore, 1950-1980.* Baltimore: Johns Hopkins University Press, 1971.

McCulloch, Rachel. "Commodity Power and the International Community." Harvard Institute of Economic Research, Discussion Paper No. 440, October 1975.

——. "Global Commodity Politics." *The Wharton Magazine,* Spring 1977.

McKern, R.B. Multinational Enterprises and Natural Resources (Sydney: McGraw-Hill, 1976).

McNicol, David L. *Commodity Agreements and Price Stabilization.* Lexington Mass.: Lexington Books, D.C. Heath, 1978.

Meade, J.E. *A Neoclassical Theory of Economic Growth.* London: Unwin Books, 1961.

Meadows, Donella H., and Meadows, Dennis. *The Limits to Growth.* New York: Universe Books, 1972.

Michaelson, D.D. "Wanted: New Systems for Surface Mining." *Engineering and Mining Journal,* October 1974.

Mikdashi, Zuhayr. *The International Politics of Natural Resources.* Ithica, N.Y.: Cornell University Press, 1976.

Mikesell, Raymond F. *Foreign Investment in Copper Mining.* Baltimore: Johns Hopkins University Press, 1975.

Miller, Jack Robert. "Iron, Steelmaking Metallics Supply Seen Meeting World Demand Forecast for '72-'85." *Engineering and Mining Journal,* September 1974.

Nordhaus, W.D. "The Allocation of Energy Resources." Brookings Institution Papers, 1973.

——. "Resources as a Constraint on Growth." *American Economic Review,* May 1974.

Park, C.F., and MacDiarmid, R.A. *Ore Deposits.* San Francisco: W.H. Freeman, 1974.

Pearce, David. *Environmental Economics.* London: Longmans, 1976.

Pearce, David, and Rose, J. *The Economics of Natural Resource Depletion.* London: Macmillan, 1975.

Petersen, Frederic M. "A Model of Mining and Exploring for Natural Resources." *Journal of Environmental Economics and Management,* no. 5, 1978.

Phillips, W.G.B., and Edwards, D.P. "Metal Prices as a Function of Ore Grade." *Resources Policy,* September 1976.

Pindyck, Robert S. "Cartel Pricing and the Structure of the World Bauxite Market." *Bell Journal of Economics,* Autumn 1977.

——. "Gains to Producers from the Cartelization of Exhaustible Resources." *The Review of Economics and Statistics,* May 1978.

——. "OPEC's Threat to the West." *Foreign Policy,* Spring 1978.

Prain, Ronald Copper: *The Anatomy of an Industry.* London: Mining Journal Books, 1975.

Quirk, James P. *Intermediate Microeconomics* Chicago: Science Research Associates, 1976.

Radetzki, Marian, and Zorn, S., *Financing Mining Projects in Developing Countries.* London: Mining Journal Books, 1979.

Radetzki, Marian. "The Potential for Monopolistic Commodity Pricing by Developing Countries." In *A World Divided: The Less Developed Countries in the International Economy,* edited by G.K. Helleiner. Cambridge: Cambridge University Press, 1975.

——. "Will the Long Run Global Supply of Industrial Minerals by Adequate?" Paper presented at the Fifth Congress of the International Economic Association, September 1977.

Rayment, Paul B.W. "On the Analysis of the Export Performance of Developing Countries." *Economic Record,* June 1971.

——. "The Homogenity of Manufacturing Industries with Respect to Factor Intensity: The Case of the United Kingdom." *Oxford Bulletin of Economics and Statistics,* August 1976.

Rogers, Paul. "The Role of Less Developed Countries in World Resource Use." In *Future Resources and World Development,* edited by Paul Rogers and Anthony Vann. New York: Plenum, 1976.

Rose, Sanford. "Third World Commodity Power Is a Costly Illusion." *Fortune,* November 1976.

Ross, Marc H., and Williams, Robert H. "Energy Efficiency: Our Most Underrated Energy Resource." *Bulletin of the Atomic Scientists,* November 1976.

Rustow, D.A., and Mugno, John. *OPEC: Success and Prospects.* New York: New York University Press, 1976.

Samuelson, Paul Anthony. *Economics: An Introductory Analysis,* 9th ed. New York: McGraw-Hill, 1975.

Seaborg, Glenn. "The Recycle Society of Tomorrow." *Futurist,* June 1974.

Singer, D.A. "Long Term Adequacy of Metal Resources." *Resource Policy,* June 1977.

Slesser, Malcolm. *Energy in the Economy.* London: Macmillan, 1978.

Smets, L., and Kovach, Y. "Metals Analysis and Outlook." *Charter Consolidated Limited.* London, 1977.

Smith, Ben. "Bilateral Monopoly and Export Price Bargaining in the Resource Goods Trade." *The Economic Record,* March 1977.

——. "Australian Minerals Development." In *Growth, Trade, and Structural*

Change in an Open Australian Economy, edited by Wolfgang Kasper and Thomas G. Parry. Sydney: Centre for Economic Research, University of New South Wales, 1978.

Smith, G., and Shink, F. "International Tin Agreement: A Reassessment." U.S. Treasury Department OASIA, Research Discussion Paper, no. 75/18, 1975.

Solow, Robert. "Richard T. Ely Lecture: The Economics of Resources or the Resources of Economics." *American Economic Review,* May 1974.

———. "Resources and Economic Growth." *The American Economist,* Fall, 1978.

Spindler, Z.A. "Endogenous Bargaining Power in Bilateral Monopoly and Bilateral Exchange," *Canadian Journal of Economics,* Aug. 1974.

Takeuchi, Kenji. "CIPEC and the Copper Earnings of Member Countries." *The Developing Countries,* February 1972.

Tilton, John. *The Future of Nonfuel Minerals.* Washington, D.C.: The Brookings Institution, 1977.

Timm, Hans J. "Kurzfristige Internationale Rohstoffpreisentwicklung und Konjunkturschwankungen," HWWA Institute für Wirtschaftsforschung Hamburg, March 1976.

Tuve, George L. *Energy, Environment, Populations and Food: Our Four Interdependent Crises."* New York: Wiley Interscience, 1976.

Van Duyne, Carl. "Commodity Cartels and the Theory of Derived Demand." *Kyklos,* 1975.

Vann, Anthony, and Rogers, Paul. *Human Ecology and World Development.* New York: Plenum, 1974.

Varon, B., and Takeuchi, Kenji. *Developing Countries and Non-Fuel Minerals. Foreign Affairs,* April 1974.

Vedavalli, R. *Market Structure of Bauxite/Alumina/Aluminum and Prospects for Developing Countries.* Washington, D.C.: The World Bank, 1977.

Warren, Kenneth. *Mineral Resources.* Harmondsworth: Penguin Books, 1973.

Vogely, William A. "Is There a Law of Demand for Minerals?" *Earth and Mineral Sciences,* vol. 45, no. 7, April 1976.

Waters, Alan Rufus. "The Economic Reason for International Commodity Agreements." *Kyklos,* no. 27, 1974.

Zettermark, Sören. "The Long Term Supply of Aluminum." Ph.D. dissertation, The University of Stockholm, 1976.

Index

About the Author

Ferdinand E. Banks is research fellow at the University of Uppsala, Sweden, and during 1978 was professorial fellow in economic policy of the Reserve Bank of Australia and visiting professor in the Department of Econometrics, the University of New South Wales (Sydney). He attentioned Illinois Institute of Technology and Roosevelt University (Chicago, Illinois), receiving the B.A. in economics. After serving with the United States Army in the Orient and Europe, he worked as an engineer and systems and procedures analyst. He received the M.Sc. and Fil. Lic. from the University of Stockholm, and he also has the Fil. Dr. from the University of Uppsala. He taught for five years at the University of Stockholm, was senior lecturer in economics and statistics at the United Nations African Institute for Economics and Development Planning, Daker, Senegal, and has been consultant lecturer in macroeconomics for the OECD in Lisbon, Portugal. From 1968 to 1971, Dr. Banks was an econometrician for the United Nations Commission on Trade and Development in Geneva, Switzerland; and he has also been a consultant on planning models and the steel industry for the United Nations Industrial Organization in Vienna. His previous books are *The World Copper Market: An Economic Analysis* (1974); *The Economics of Natural Resources* (1976); *Scarcity, Energy, and Economic Progress* (1977); and *The International Economy: A Modern Approach* (1979). He has also published 45 articles and notes in various journals and collections of essays.